DISASTER MANAGEMENT

DISASTER MANAGEMENT

Dr. S.R. Singh

A P H PUBLISHING CORPORATION
4435-36/7, ANSARI ROAD, DARYA GANJ
NEW DELHI-110 002

Published by
S.B. Nangia
A P H Publishing Corporation
4435-36/7, Ansari Road, Darya Ganj
New Delhi-110002
☎ 23274050
Email : aphbooks@vsnl.net

2026

Typesetting at
Paragon Computers
B-36, Chanakya Place

Printed at
Balaji Offset
Navin Shahdara, Delhi-32
Ph.: 22324437

Preface

Whether it be natural as in the forms of earthquakes, landslides and volcanic eruptions or man-made like in the case of fires bomb blasts and wars, disasters have always been a part and parcel of human civilisation. One of the biggest and most tragic of disasters which occurred in recent history, was the 2004 South-Asian tsunami disaster, which wreaked havoc in many of the Asian countries, resulting in immense loss of life and property. Heralding tragedy in nearly every sphere of life, disasters are nevertheless unpredictable, and quite often, beyond control. However, considering the price paid in cases of disasters, one can always make sure that preparedness is always maintained in unforeseen circumstances, and that post-disaster management is deftly handled.

Disaster management entails the process of mitigation, response and recovery, the co-ordinated efforts of all three aspects being crucial to the way to make sure disasters are handled with efficiency and skill. The book undertakes a thorough, pragmatic approach to the issue, bringing under cover both pre- as well as post-disaster management techniques and strategies. Exploring a wide range of matters within the subject, it provides the readers with content which is aimed at providing practical solutions to concerns. In addition, the book delineates current trends and practices in the field, apart from providing case studies and examples, making the book more relevant and up-to-date.

Dr. S.R. Singh

Contents

1

Introduction

Disasters occur rapidly, instantaneously and indiscriminately. These extreme events either natural or man-induced exceed the tolerable magnitude within or beyond certain time limits, make adjustment difficult, result in catastrophic losses of property and income and life is paralysed. This events which occur aggravate natural environmental processes to cause disasters to human society such as sudden tectonic movements leading to earthquake and volcanic eruptions, continued dry conditions leading to prolonged droughts, floods, atmospheric disturbances, collision of celestial bodies etc. It may be mentioned that environmental disasters are always viewed in terms of human beings. The intensity of environmental disaster is weighed in terms of the quantum of damage done to the human society.

Hazardous environmental processes always create extreme events but not all the extreme events become disasters. These may become disasters only when they adversely affect human society. For example, a very strong tropical cyclone(typhoon, hurricane or tornado) becomes only extreme event when it occurs and dies in the midst of an ocean but it becomes disaster when it strikes the inhabited coastal areas and inflicts colossal loss to human property and life. Similarly, a volcanic eruption in uninhabited land or ocean is never disastrous but when it takes place in densely populated area, it becomes disaster.

Generally, the environmental disasters are natural and hence, these are termed as natural disasters. In other words, the natural sudden physical process and events become disasters when people live close to a potential danger. For example, if an earthquake of more than 10 on Richter scale occurs in totally uninhabited area it is not a disaster at all but an earthquake event of lower intensity, say below 7 on Richter scale, occurs in heavily populated area, it becomes a severe disaster. It may be further pointed out that it is not the frequency which makes any extreme event disastrous rather it is intensity, magnitude and dimension and the quantum of damage done by any event which makes it disastrous.

1.1. Types of Disasters

Environmental disaster are normally divided into two broad categories on the basis of main causative factors viz.

— Natural disasters and

— Man-induced disasters

Natural disasters are further sub-divided into two categories e.g. (1) Planetary disasters and (2) Extra-terrestrial or Extra-planetary disasters. Planetary disaster again fall into two sub types viz. (a) Terrestrial or Endogenous disasters and (b) Atmospheric or Exogenous disasters.

Man-induced disasters may be divided into three sub-categories viz. (a) Physical man-induced disasters like landslides, accelerated soil erosion (b) Chemical and nuclear disaster like release of toxic chemical in the air, nuclear explosions, leakage of radioactive elements and (c) Biological disasters like sudden increase or decrease of population of species in a given habitat either due to increased nutrients or increase of toxic chemical elements.

1.1.1. Natural Disasters

A natural hazard has an element of human involvement. A physical event, such as a volcanic eruption, that does not affect human beings is a natural phenomenon but not a natural hazard. A natural phenomenon that occurs in a populated area is a hazardous event. A hazardous event that causes unacceptably large numbers of fatalities and/ or overwhelming property damage is a natural disaster. In areas where there are no human interests, natural phenomena do not constitute hazards nor do they result in disasters.

Although humans can do little or nothing to change the incidence or intensity of most natural phenomena, they have an important role to play in ensuring that natural events are not converted into disasters by their own actions. It is important to understand that human intervention can increase the frequency and severity of natural hazards. For example, when the toe of a landslide is removed to make room for a settlement, the earth can move again and bury the settlement. Human intervention may also cause natural hazards where none existed before.

Volcanoes erupt periodically, but it is not until the rich soils formed on their eject are occupied by farms and human settlements that they are considered hazardous. Finally, human intervention reduces the mitigating effect of natural ecosystems. Destruction of coral reefs, which removes the shore's first line of defense against ocean currents and storm surges, is a clear example of an intervention that diminishes the ability of an ecosystem to protect itself. An extreme case of destructive human intervention into an ecosystem is desertification, which, by its very definition, is a human-induced "natural" hazard.

All this is the key to developing effective vulnerability reduction measures: if human activities can cause or aggravate the destructive effects of natural phenomena, they can also eliminate or reduce them.

1.1.1.1. Natural hazards and Sustainable Development

In a general sense, these tasks may be called "environmental planning"; they consist of diagnosing the needs of an area and identifying the resources available to it, then using this information to formulate an integrated development strategy composed of sectoral investment projects. This process uses methods of systems analysis and conflict management to arrive at an equitable distribution of costs and benefits, and in doing so it links the quality of human life to environmental quality. In the planning work, then, the environment-the structure and function of the ecosystems that surround and support human life-represents the conceptual framework. In the context of economic development, the environment is that composite of goods, services, and constraints offered by surrounding ecosystems.

An ecosystem is a coherent set of interlocking relationships between and among living things and their environments. For example, a forest is an ecosystem that offers goods, including trees that provide lumber, fuel, and fruit. The forest may also provide services in the form of water storage and flood control, wildlife habitat, nutrient storage, and recreation. The forest, however, like any physical resource, also has its constraints. It requires a fixed period of time in which to reproduce itself, and it is vulnerable to wildfires and blights. These vulnerabilities, or natural hazards, constrain the development potential of the forest ecosystem.

Earthquakes

Earthquakes are caused by the sudden release of slowly accumulated strain energy along a fault in the earth's crust, Earthquakes and volcanoes occur most commonly at the collision zone between tectonic plates. Earthquakes represent a particularly severe threat due to the irregular time intervals between events, lack of adequate forecasting, and the hazards associated with these:

— Ground shaking is a direct hazard to any structure located near the earthquake's center. Structural failure takes many human lives in densely populated areas.

— Faulting, or breaches of the surface material, occurs as the separation of bedrock along lines of weakness.

— Landslides occur because of ground shaking in areas having relatively steep topography and poor slope stability.

— Liquefaction of gently sloping unconsolidated material can be triggered by ground shaking. Rows and lateral spreads (liquefaction phenomena) are among the most destructive geologic hazards.

— Subsidence or surface depressions result from the settling of loose or unconsolidated sediment. Subsidence occurs in waterlogged soils, fill, alluvium, and other materials that are prone to settle.

— Tsunamis or seismic sea waves, usually generated by seismic activity under the ocean floor, cause flooding in coastal areas and can affect areas thousands of kilometers from the earthquake center.

Volcanoes

Volcanoes are perforations in the earth's crust through which molten rock and gases escape to the surface. Volcanic hazards stem from two classes of eruptions:

— Explosive eruptions which originate in the rapid dissolution and expansion of gas from the molten rock as it nears the earth's surface. Explosions pose a risk by scattering rock blocks, fragments, and lava at varying distances from the source.

— Effusive eruptions where material flow rather than explosions is the major hazard. Flows vary in nature (mud, ash, lava) and quantity and may originate from multiple sources. Flows are governed by gravity, surrounding topography, and material viscosity.

Hazards associated with volcanic eruptions include lava flows, falling ash and projectiles, mudflows, and toxic gases. Volcanic activity may also trigger other natural hazardous events including local tsunamis, deformation of the landscape, floods when lakes are breached or when streams and rivers are dammed, and tremor-provoked landslides.

Landslides

The term landslide includes slides, falls, and flows of unconsolidated materials. Landslides can be triggered by earthquakes, volcanic eruptions, soils saturated by heavy rain or groundwater rise, and river undercutting. Earthquake shaking of saturated soils creates particularly dangerous conditions. Although landslides are highly localised, they can be particularly hazardous due to their frequency of occurrence. Classes of landslide include:

— Rockfalls, which are characterised by free-falling rocks from overlying cliffs. These often collect at the cliff base in the form of talus slopes which may pose an additional risk.

— Slides and avalanches, a displacement of overburden due to shear failure along a structural feature. If the displacement occurs in surface material without total deformation it is called a slump.

— Flows and lateral spreads, which occur in recent unconsolidated material associated with a shallow water table. Although associated with gentle topography, these liquefaction phenomena can travel significant distances from their origin.

The impact of these events depends on the specific nature of the landslide. Rockfalls are obvious dangers to life and property but, in general, they pose only a localised threat due to their limited areal influence. In contrast, slides, avalanches, flows, and lateral spreads, often having great areal extent, can result in massive loss of lives and property. Mudflows, associated with volcanic eruptions, can travel at great speed from their point of origin and are one of the most destructive volcanic hazards.

A survey of environmental constraints, whether focused on urban, rural, or wildland ecosystems, includes (1) the nature and severity of resource degradation; (2) the underlying causes of the degradation, which include the impact of both natural phenomena and human use; and (3) the range of feasible economic, social, institutional, policy, and financial interventions designed to retard or alleviate degradation. In this sense, too, natural hazards must be considered an integral aspect of the development planning process.

Recent development literature sometimes makes a distinction between "environmental projects" and "development projects." "Environmental projects" include objectives such as sanitation, reforestation, and flood control, while "development projects" may focus on potable water supplies, forestry, and irrigation. But the project-by-project approach is clearly an ineffective means of promoting socioeconomic well-being. Development projects, if they are to be sustainable, must incorporate sound environmental management. By definition, this means that they must be designed to improve the quality of life and to protect or restore environmental quality at the same time and must also ensure that resources will not be degraded and that the threat of natural hazards will not be exacerbated. In short, good natural hazard management is good development project management.

Indeed, in high-risk areas, sustainable development is only possible to the degree that development planning decisions, in both the public and private sectors, address the destructive potential of natural hazards. This approach is particularly relevant in post-disaster situations, when tremendous pressures are brought to bear on local, national, and international agencies to replace, frequently on the same site, destroyed facilities. It is at such times that the pressing need for natural hazard and risk assessment information and its incorporation into the development planning process become most evident.

To address hazard management, specific action must be incorporated into the various stages of the integrated development planning study: first, an assessment of the presence and effect of natural events on the goods and services provided by natural resources in the plan area; second, estimates of the potential impact of natural events on development activities; and third, the inclusion of measures to reduce vulnerability in the proposed development activities. Within this framework, "lifeline" networks should be identified: components or critical segments of production facilities, infrastructure, and support systems for human settlements, which should be as nearly invulnerable as possible and be recognised as priority elements for rehabilitation following a disaster.

Flooding

Two types of flooding can be distinguished: (1) land-borne floods, or river flooding, caused by excessive run-off brought on by heavy rains, and (2) sea-borne floods, or coastal flooding, caused by storm surges, often exacerbated by storm run-off from the upper watershed. Tsunamis are a special type of sea-borne flood.

Coastal flooding

Storm surges are an abnormal rise in sea water level associated with hurricanes and other storms at sea. Surges result from strong on-shore winds and/or intense low pressure cells and ocean storms. Water level is controlled by wind, atmospheric pressure, existing astronomical tide, waves and swell, local coastal topography and bathymetry, and the storm's proximity to the coast.

Most often, destruction by storm surge is attributable to:

- Wave impact and the physical shock on objects associated with the passing of the wave front.
- Hydrostatic/dynamic forces and the effects of water lifting and carrying objects.

The most significant damage often results from the direct Impact of waves on fixed structures. Indirect impacts include flooding and undermining of major infrastructure such as highways and railroads. Hooding of deltas and other low-lying coastal areas is exacerbated by the influence of tidal action, storm waves, and frequent channel shifts.

River flooding

Land-borne floods occur when the capacity of stream channels to conduct wafer is exceeded and water overflows banks. Floods are natural phenomena, and may be expected to occur at irregular intervals on all stream and rivers. Settlement of floodplain areas is a major cause of flood damage.

Tsunamis

Tsunamis are long-period waves generated by disturbances such as earthquakes, volcanic activity, and undersea landslides. The crests of these waves can exceed heights of 25 meters on reaching shallow water. The unique characteristics of tsunamis (wave lengths commonly exceeding 100 km, deep-ocean velocities of up to 700 km/hour, and small crest heights in deep water) make their detection and monitoring difficult. Characteristics of coastal flooding caused by tsunamis are the same as those of storm surges.

Hurricanes

Hurricanes are tropical depressions which develop into severe storms characterised by winds directed inward in a spiraling pattern toward the center. They are generated over warm ocean water at low latitudes and are particularly dangerous due to their destructive potential, large zone of influence, spontaneous generation, and erratic movement. Phenomena which are associated with hurricanes are:

— Winds exceeding 64 knots (74 mi/hr or 118 km/hr), the definition of hurricane force. Damage results from the wind's direct impact on fixed structures and from wind-borne objects.

— Heavy rainfall which commonly precedes and follows hurricanes for up to several days. The quantity of rainfall is dependent on the amount of moisture in the air, the speed of the hurricane's movement, and its size. On land, heavy rainfall can saturate soils and cause flooding because of excess runoff (land-borne flooding); it can cause landslides because of added weight and lubrication of surface material; and/or it can damage crops by weakening support for the roots.

— Storm surge, which, especially when combined with high tides, can easily flood low-lying areas that are not protected.

1.1.1.2. Mitigation of natural hazards

In the field of landslide mitigation, a study in the State of New York (U.S.A.) showed that improved procedures from 1969 to 1975 reduced the cost of repairing landslide damage to highways by over 90 percent. Experience of the city of Los Angeles, California, indicates that adequate grading and soil analysis ordinances can reduce landslide losses by 97 percent.

A study in the San Fernando Valley, California, after the 1971 earthquake showed that of 568 older school buildings that did not satisfy the requirements of the Field Act (a law stipulating design standards), 50 were so badly damaged that they had to be demolished. But all of the 500 school buildings that met seismic-resistance standards suffered no structural damage. The Loma Prieta earthquake in 1989 was the costliest natural disaster in U.S. history, but provisions in local zoning and building codes kept it from being even worse. In the San Francisco Bay area post-1960 structures swayed but stayed intact, while older buildings did not fare nearly as well. Unreinforced masonry structures suffered the worst damage. Buildings on solid ground were less likely to sustain damage than those constructed on landfill or soft mountain slopes.

Mitigation techniques can also lengthen the warning period before a volcanic eruption, making possible the safe evacuation of the population at risk. Sensitive monitoring devices can now detect increasing volcanic activity months in advance of an eruption. Still more sophisticated assessment, monitoring, and alert systems are becoming available for volcanic eruption, hurricane, tsunami, and earthquake hazards.

Sectoral hazard assessments conducted by the OAS of, among others, energy in Costa Rica and agriculture in Ecuador have demonstrated the savings in capital and continued production that can be realised with very modest investments in the mitigation of natural hazard threats through vulnerability reduction and better sectoral planning.

1.1.2. Hazards in Arid and Semi-Arid Areas

1.1.2.1. Desertification

Desertification, or resource degradation in arid lands that creates desert conditions, results from interrelated and interdependent sets of actions, usually brought on by drought combined with human and animal population pressure. Droughts are prolonged dry periods in natural climatic cycles. The cycles of dry and wet periods pose serious problems for pastoralists and farmers who gamble on these cycles. During wet periods, the sizes of herds are increased and cultivation is extended into drier areas. Later, drought destroys human activities which have been extended beyond the limits of a region's carrying capacity.

Overgrazing Is a frequent practice In dry lands and is the single activity that most contributes to desertification. Dry-land farming refers to rain-fed agriculture In semiarid regions where water is the principal factor limiting crop production. Grains and cereals are the most frequently grown crops. The nature of dry-land farming makes it a hazardous practice which can only succeed if special conservation measures such as stubble mulching, summer fallow, strip cropping, and clean tillage are followed. Desertified dry lands in Latin America can usually be attributed to some combination of exploitative land management and natural climate fluctuations.

1.1.2.2. Erosion and Sedimentation

Soil erosion and the resulting sedimentation constitute major natural hazards that produce social and economic losses of great consequence. Erosion occurs in all climatic conditions, but is discussed as an arid zone hazard because together with salinisation, it is a major proximate cause of desertification. Erosion by water or wind occurs on any sloping land regardless of its use. Land uses which increase the risk of soil erosion Include overgrazing, burning and/or exploitation of forests, certain agricultural practices, roads and trails, and urban development. Soil erosion has three major effects: loss of support and nutrients necessary for plant growth; downstream damage from sediments generated by erosion; and depletion of the water storage capacity, because of soil loss and sedimentation of streams and reservoirs, which results in reduced natural stream flow regulation.

Stream and reservoir sedimentation is often the root of many water management problems. Sediment movement and subsequent deposition in reservoirs and river beds reduces the useful lives of water storage reservoirs, aggravates flood water damage, impedes navigation, degrades water quality, damages crops and infrastructure, and results in excessive wear of turbines and pumps.

1.1.2.3. Salinisation

Saline water is common in dry regions, and soils derived from chemically weathered marine deposits (such as shale) are often saline. Usually, however, saline soils have received salts transported by water from other locations. Salinisation most often occurs on irrigated land as the result of poor water control, and the primary source of salts impacting soils is surface and/or ground water. Salts accumulate because of flooding of low-lying lands, evaporation from depressions having no outlets, and the rise of ground water close to soil surfaces. Salinisation results in a decline in soil fertility or even a total loss of land for agricultural purposes. In certain instances, farmland abandoned because of salinity problems may be subjected to water and wind erosion and become desertified.

Inexpensive water usually results in over-watering. In dry regions, salt-bearing ground water is frequently the major water resource. The failure to properly price water from irrigation projects can create a great demand for such projects and result in misuse of available water, causing waterlogging and salinisation.

However, much remains to be done. The overall record of hazard management in Latin America and the Caribbean is unimpressive for a number of reasons-among them lack of awareness of the issue, lack of political incentive, and a sense of fatalism about "natural" disasters. But techniques are becoming available, experiences are being analysed and transmitted, the developing countries have demonstrated their interest, and the lending agencies are discussing their support. If these favorable tendencies can be encouraged, significant reduction of the devastating effects of hazards on development in Latin America and the Caribbean is within reach.

1.1.2.4. Susceptibility to Vulnerability Reduction

Rapid onset vs. slow onset

The speed of onset of a hazard is an important variable

since it conditions warning time. At one extreme earthquakes, landslides, and flash floods give virtually no warning. Less extreme are tsunamis, which typically have warning periods of minutes or hours, and hurricanes and floods, where the likelihood of occurrence is known for several hours or days in advance. Volcanoes can erupt suddenly and surprisingly, but usually give indications of an eruption weeks or months in advance. Other hazards such as drought, desertification, and subsidence act slowly over a period of months or years. Hazards such as erosion/sedimentation have varying lead times: damage may occur suddenly as the result of a storm or may develop over many years.

Controllable events vs. immutable events

For some types of hazards the actual dimensions of the occurrence may be altered if appropriate measures are taken. For others, no known technology can effectively alter the occurrence itself. For example, channelising a stream bed can reduce the areal extent of inundations, but nothing will moderate the ground shaking produced by an earthquake.

Frequency vs. severity

Where flooding occurs every year or every few years, the hazard becomes part of the landscape, and projects are sited and designed with this constraint in mind. Conversely, in an area where a tsunami may strike any time in the next 50 or 100 years, it is difficult to stimulate interest in vulnerability reduction measures even though the damage may be catastrophic. With so long a time horizon, investment in capital intensive measures may not be economically viable. Rare or low-probability events of great severity are the most difficult to mitigate, and vulnerability reduction may demand risk-aversion measures beyond those justified by economic analysis.

Mitigation measures to withstand impact vs. mitigation measures to avoid impact

Earthquake-resistant construction and floodproofing of buildings are examples of measures that can increase the capacity of facilities to withstand the impact of a natural hazard. Measures such as zoning ordinances, insurance, and tax incentives, which direct uses away from hazard-prone areas, lead to impact avoidance. Mitigation measures are both more critically needed and more amenable to economic justification than in less-developed areas. Urban areas are likely to have or are able to establish the institutional arrangements necessary for hazard management.

For small towns and villages non-structural mitigation measures may be the only affordable alternative. Such settlements rely on the government to only a limited extent for warning of an impending hazard or assistance in dealing with it. Thus organising the local community to cope with hazards is a special aspect of hazard management. The physical characteristics of the land, land-use patterns, susceptibility to particular hazards, income level, and cultural characteristics similarly condition the options of an area in dealing with natural hazards.

1.1.3. Hazard Management and Development Planning

Development planning is considered the process by which governments produce plans-consisting of policies, projects, and supporting actions-to guide economic, social, and spatial development over a period of time. The hazard management process consists of a number of activities designed to reduce loss of life and destruction of property. Natural hazard management has often been conducted independently of development planning.

1.1.3.1. Hazard management activities

The natural hazard management process can be divided into pre-event measures, actions during and immediately following an event, and post-disaster measures. In approximate chronological order these are as follows:

1. Pre-event Measures:
   - a. Mitigation of Natural Hazards:
     - Data Collection and Analysis
     - Vulnerability Reduction
   - b. Preparation for Natural Disasters
     - Prediction
     - Emergency Preparedness (including monitoring, alert, evacuation)
     - Education and Training
2. Measures During and Immediately After Natural Disasters:
   - a. Rescue
   - b. Relief
3. Post-disaster Measures:
   - a. Rehabilitation
   - b. Reconstruction

Disaster mitigation

An accurate and timely prediction of a hazardous event can save human lives but does little to reduce economic losses or social disruption; that can only be accomplished by measures taken longer in advance. Included in the concept of disaster mitigation is the basic assumption that the impact of disasters can be avoided or reduced when they have been anticipated during development planning. Mitigation of disasters usually entails reducing the vulnerability of the elements at risk, modifying the hazard-proneness of the site, or changing its function. Mitigation measures can have a structural character, such as the inclusion of specific safety or vulnerability reduction measures in the design and construction of new facilities, the retrofitting of existing facilities, or the building of protective devices. Non-structural mitigation measures typically concentrate on limiting land uses, use of tax incentives and eminent domain, and risk underwriting through insurance programs.

Many countries are making efforts to introduce mitigation measures in hazard-prone areas. For example, the coastal area of Ecuador and the northern area of Peru are often affected by severe floods caused by "El Niño" or the El Niño Southern Oscillation (ENSO) phenomenon, which recurs approximately every 3 to 16 years. Between November 1982 and June 1983, heavy rains created the most dramatic series of floods reported this century, affecting 12,000 square kilometers in this region, with total losses estimated at US$1,200 million. Subsequently, Peru transferred six of the most affected villages to higher elevations (a non-structural mitigation measure), and introduced special adobe-building techniques to strengthen new constructions against earthquakes and floods (a structural mitigation measure).

Disaster mitigation also includes the data collection and analysis required to identify and evaluate appropriate measures and include them in development planning. The data collection involves essentially three kinds of studies:

1.1.4. Natural Hazard Assessments

Studies that assess hazards provide information on the probable location and severity of dangerous natural phenomena and the likelihood of their occurring within a specific time period in a given area. These studies rely heavily on available scientific information, including geologic, geomorphic, and soil maps; climate and hydrological data; and topographic maps, aerial photographs, and satellite imagery. Historical information, both written reports and oral accounts from long-term residents, also helps characterise potential hazardous events. Ideally, a natural hazard assessment promotes an awareness of the issue in a developing region, evaluates the threat of natural hazards, identifies the additional information needed for a definitive evoluation, and recommends appropriate means of obtaining it.

1.1.4.1. Vulnerability assessments

Vulnerability studies estimate the degree of loss or damage that would result from the occurrence of a natural phenomenon of given severity. The elements analysed include human population/capital facilities and resources such as settlements, lifelines, production facilities, public assembly facilities, and cultural patrimony; and economic activities and the normal functioning of settlements.

Vulnerability can be estimated for selected geographic areas, e.g., areas with the greatest development potential or already developed areas in hazardous zones. The techniques employed include lifeline (or critical facilities) mapping and sectoral vulnerability analyses for sectors such as energy, transport, agriculture, tourism, and housing. In Latin America and the Caribbean vulnerability to natural hazards is rarely considered in evaluating an investment even though vulnerability to other risks, such as fluctuating market prices and raw-material costs, is taken into account as standard practice.

1.1.4.2. Risk assessments

Information from the analysis of an area's hazards and its

vulnerability to them is integrated in an analysis of risk, which is an estimate of the probability of expected loss for a given hazardous event. Formal risk analyses are time-consuming and costly, but shortcut methods are available which give adequate results for project evaluation. Once risks are assessed, planners have the basis for incorporating mitigation measures into the design of investment projects and for comparing project versus no-project costs and benefits.

1.1.5. Natural Hazard Prediction

Even short notice of the probable occurrence and effects of a natural phenomenon is of great importance in reducing loss of life and property. The prediction of a natural event is a direct outcome of scientific investigation into its causes and is aimed at establishing the probability of the next occurrence in terms of time, place, and range of severity. Increasingly sophisticated monitoring stations, both manned and remote, collect information of potentially hazardous events for more accurate prediction.

Some hazards, such as hurricanes and floods, can be forecast with high accuracy, but most geologic events cannot. Alert systems for some kinds of disasters suffer from a very short lead time. In the case of tsunamis, for example, the Pacific Warning Center, which constantly monitors the oceans, provides advance notice that varies from ten minutes to a few hours. At best, these warnings provide enough time to withdraw the population, but not to take other preventive measures.

1.1.6. Man-induced Disasters

Disasters having an element of human intent, negligence, error or the ones involving the failure of a system are called man-made disasters. Man-made hazards are in turn categorised as technological or sociological. Technological hazards are results of failure of technology, such as engineering failures, transport accidents or environmental disasters. Sociological hazards have a strong human motive, such as crime, stampedes, riots and war.

1.1.7. Vulnerable Elements in the Planning Process

1.1.7.1. Human settlements

Human population and associated housing and services.

Critical facilities

(1) Essential services, such as telecommunications, water, energy, and sanitation; (2) emergency medical services, fire and police stations, and disaster organisations; and (3) local, national, and international transportation facilities and carriers.

Economic production facilities

Major sources of livelihood of the population, such as industries, banking and commerce buildings, public markets, agroprocessing plants and areas of agricultural production, livestock, forestry, mines, and fisheries production.

Public assembly sites

Buildings such as schools, churches, auditoriums, theatres, public markets, and public and private office buildings.

Cultural patrimony

Buildings of significant cultural and community value or use, and buildings of architectural importance.

Although world-wide efforts to anticipate earthquakes persist, their prediction is still an incipient science. Few forewarnings have been as successful as the one made in February 1975 when the people of Haicheng, China, were evacuated six hours before a magnitude M.7 earthquake struck. Other predictions have been disastrous, as was the case with the erroneous warning of an imminent earthquake in Peru in 1981. Thousands of people fled, causing some deaths and long-term disruption of investment and tourism.

1.1.8. Emergency Preparedness

Emergency preparedness is aimed at minimising the loss of life and property during a natural event. Preparedness includes actions taken in anticipation of the event and special activities both during and immediately after the event.

Two levels of preparedness can be identified: public safety information and hazard awareness planning. The first includes a number of efforts aimed at increasing the amount of information disseminated to the public and at promoting cooperation between the public and the authorities in case of an emergency. In the course of an event, or in its aftermath, social and public behavior undergoes important changes. This results in new organisational responsibilities for the public sector. Hazard information and education programs can improve public preparedness and social conduct during a disaster.

Hazard awareness planning is concerned about improving the ability of a particular area, region, or nation to respond to natural disasters. Disaster preparedness promotes the development of a system for monitoring known hazards, a warning system, emergency and evacuation plans, emergency routes, and the formulation of educational programs for public officials and professionals. Many Latin America and Caribbean countries are developing and adopting emergency plans in order to identify and effectively mobilise human and national resources in case of a disaster.

1.1.8.1. Disaster rescue and relief

After a natural calamity, local residents usually undertake the first relief activities. However, their efforts must usually be complemented with those of national or regional authorities. The keystones of post-disaster relief are the preparation of lifelines or critical facilities for emergency response, training, disaster rehearsals, and the identification and allocation of local and external resources.

Relief activities are affected by broad-scale planning decisions, but they are not a part of the mainstream national and regional planning processes. Although relief and disaster preparedness receive the most resources at the international, national, regional, and local levels, cost-effective mitigation measures are not adequately considered. This lack of forethought exacerbates the effects of natural disasters in terms of loss of life and property. Meanwhile, natural disasters continue to occur worldwide, and the number of people affected is increasing faster than the population growth rate.

1.1.8.2. Post-disaster rehabilitation and reconstruction

Concurrent with or immediately after relief activities, post-disaster rehabilitation is carried out to restore the normal functions of public services, business, and commerce, to repair housing and other structures, and to return production facilities to operation. However, mitigation is often ignored in this phase: rehabilitation proceeds without any measures to reduce the chances of the same impact if the event happens again. In developing countries, road systems that are flooded or blocked by landslides year after year are commonly rebuilt at the same site and with similar design specifications.

In considering reconstruction costs, existing development policies and sectoral projects need to be reevaluated. In many cases, they are no longer appropriate or do not coincide with the best use of natural resources. For this reason, the natural hazard management process must examine any changes in the resources, goals, objectives, and products of development plans and incorporate these factors into subsequent planning activities.

1.1.8.3. Education and training activities

Education and training, both formal and informal, prepare people at all levels to participate in hazard management. Universities, research centers, and international development assistance agencies play the leading formal role in preparing individuals in a variety of skill levels such as natural hazards assessment, risk reduction, and natural phenomena prediction. These activities are also carried out by operational entities such as ministries of agriculture, transportation, public works, and defense.

Informal learning can be delivered through brochures, booklets, and audio and video tapes prepared by national and international agencies involved in disaster preparedness and mitigation programs, and through the national media. Additionally, courses, workshops, conferences, and seminars organised by national and international disaster assistance agencies disseminate great amounts of information on natural hazard management strategies.

Finally, direct observation after a disaster has proved to be one of the most effective means of learning. Post-disaster investigations describe the qualitative and quantitative aspects of natural hazards, often improving on information produced by modelling and conjecture by indicating areas where development should be extremely limited or should not take place.

A direct outcome of the learning process is (1) the improvement of policies and program actions, building codes, standards, construction and design skills; (2) the development of legislation to mandate the adoption of these policies and the strengthening or creation of new disaster organisations; (3) the improvement of the key logistical aspects of disaster prevention, such as communication and warning systems; and (4) the establishment of community and resource organisations to confront future disasters.

1.1.8.4. Integrated development planning

Integrated development planning is a multidisciplinary, multisectoral approach to planning. Issues in the relevant economic and social sectors are brought together and analyzed vis-a-vis the needs of the population and the problems and opportunities of the associated natural resource base. A key element of this process is the generation of investment projects, defined as an investment of capital to create assets capable of generating a stream of benefits over time. A project may be independent or part of a package of projects comprising an integrated development effort. The process of generating projects is called the project cycle. This process proceeds from the establishment of development policies and strategies, the identification of project ideas, and the preparation of project profiles through prefeasibility and feasibility analyses (and, for large projects, design studies) to final project approval, financing, implementation, and operation.

While the process is more or less standardised, each agency develops its own version. The development planning process consists of four stages: Preliminary Mission, Phase I (development diagnosis), Phase II (project formulation and preparation of an action plan), and Implementation. Because the process is cyclical, activities relating to more than one stage can take place at the same time.

Generally, planners depend on the science and engineering community to provide the required information for natural hazard assessments. If the information available is adequate, the planner may decide to make an assessment. If it is not adequate, the planner usually decides that the time and cost of generating more would be excessive, and the assessment is not made. While the information available on hurricanes and geologic hazards is often adequate for a preliminary evaluation, the information on desertification, flooding, and landslide hazards rarely is. The OAS has developed fast, low-cost methodologies that make these evaluations possible in the context of a development study.

Preliminary mission

The first step in the process of technical assistance for an integrated development planning study is to send a "preliminary mission" to consult with officials in the interested country. Experience has shown that this joint effort of OAS staff and local planners and decision-makers is frequently the most critical event in the entire study. They take action to:

— Determine whether the study area is affected by one or more natural hazards. For example, the National Environmental Study of Uruguay conducted by the

OAS with financial support from the Inter-American Development Bank determined in the preliminary mission that natural hazards were an important environmental problem, and consequently an assessment of all significant hazards, to be conducted by reviewing existing information, was programmed for Phase I.

— Identify the information available for judging the threat posed by those hazards in the study area: history of hazardous events; disaster and damage reports; assessments of hazards, vulnerability, risk; maps and reports on natural resources and hazards; topographic maps, aerial photographs, satellite imagery.

— Determine whether the available data are sufficient to evaluate the threat of hazards. If they are not, determine what additional data collection, hazard assessment, remote sensing, or specialised equipment will be needed for the next stage of the study. For example, in preliminary missions in Dominica, Saint Lucia, and St. Vincent and the Grenadines, landslides were determined to be a serious problem, and landslide assessments were included in the work plan for Phase I.

— Determine whether the studies required would serve more than one sector or project. If so, establish coordination.

— Establish coordination with the national institution responsible for disaster planning.

— Prepare an integrated work plan for Phase I that specifies the hazard work to be done, the expertise needed, and the time and cost requirements.

Phase I: development diagnosis

In Phase I, the team analyses the study region and arrives

at detailed estimates of development potentials and problems of the region and selected target areas. From this analysis a multisectoral development strategy and a set of project profiles are prepared for review by government decision-makers. Phase I also includes a detailed assessment of natural hazards and the elements at risk in highly vulnerable areas which facilitates the early introduction of non-structural mitigation measures. During this phase the team will:

— Prepare a base map.

— Determine the goods, services, and hazards of the region's ecosystems. Identify cause-and-effect relationships between natural events and between natural events and human activity. In the hilly Chixoy region of Guatemala, for example, it was found that inappropriate road construction methods were causing landslides and that landslides, in turn, were the main problem of road maintenance. In Ecuador, the discovery that most of the infrastructure planned for the Manabí Water Development Project was located in one of the country's most active earthquake zones prompted a major reorientation of the project.

— Evaluate socioeconomic conditions and institutional capacity. Determine the important linkages between the study region and neighboring regions.

— Delineate target areas of high development potential, followed by more detailed natural resource and socioeconomic studies of these areas.

— In planning the development of multinational river basins or border areas where a natural disaster could precipitate an international dispute, make an overall hazard assessment as part of the resource evaluation. Examples of such studies include those for the development of the San Miguel-Putumayo River Basin, conducted in support of the Colombia-Ecuador

Joint Commission of the Amazon Cooperation Project, and for the Dominican Republic and Haiti Frontier Development Projects.

— Conduct assessments of natural hazards determined to be a significant threat in the study region. For hurricanes and geologic hazards, the existing information will probably suffice; if the information on geologic hazards is inadequate, an outside agency should be asked to conduct an analysis. For flooding, landslides, and desertification, the planning team itself should be able to supplement the existing information and prepare analyses. The studies of the Honduran departments of Atlántida and Islas de la Bahía included flood hazard assessment as part of the coastal area development plan and landslide hazard assessments for some of the inland areas.

— Conduct vulnerability studies for specific hazards and economic sectors. Prepare lifeline maps, hazard zoning studies, and multiple hazard maps as required. The study of the vulnerability of the Ecuadorian agriculture sector to natural hazards and of ways to reduce the vulnerability of lifelines in St. Kitts and Nevis, for example, both generated project ideas which could be studied at the prefeasibility level in Phase II. The study of the Paraguayan Chaco included flood and desertification assessments and multiple-hazard zoning. The execution of these hazard-related activities did not distort the time or cost of the development diagnosis.

— Identify hazard-prone areas where intensive use should be avoided.

— Prepare a development strategy, including non-structural mitigation measures as appropriate.

— Identify project ideas and prepare project profiles that address the problems and opportunities and that are compatible with political, economic, and institutional constraints and with the resources and time frame of the study.

— Identify structural mitigation measures that should be incorporated into existing facilities and proposed projects.

— Prepare an integrated work plan for the next stage that includes hazard considerations.

Phase II: Project formulation and action plan preparation

At the end of Phase I a development strategy and a set of project profiles are submitted to the government. Phase II begins after the government decides which projects merit further study. The team now makes prefeasibility and feasibility analyses of the projects selected. Refined estimates are made of benefits (income stream, increases in production, generation of employment, etc.) and costs (construction, operation and maintenance, depletion of resources, pollution effects, etc.). Valuative criteria are applied, including net present value, internal rate of return, cost-benefit ratio, and repayment possibilities. Finally, the team assembles packages of investment projects for priority areas and prepares an action plan. More detail on this phase is given in the section on Hazard Mitigation Strategies for Development Projects, but broadly speaking the team must:

— Examine the human activities that could contribute to natural hazards (e.g., irrigation, plowing in the dry season, and animal husbandry could cause or exacerbate desertification) and the social and cultural factors that could influence project vulnerability during and after implementation.

— Determine the levels of technology, credit, knowledge, information, marketing, etc., that it is realistic to expect will be available to the users of the land, and ensure that the projects formulated are based on these levels.

— Prepare site-specific vulnerability and risk assessments and appropriate vulnerability reduction measures for all projects being formulated. For example, the multimillion-dollar program for the development of the metropolitan area of Tegucigalpa, Honduras, featured landslide mitigation components. Flood alert and control projects were central elements in the comprehensive Water Resource Management and Flood Disaster Reconstruction Project for Alagoas, Brazil.

— Mitigate the undesirable effects of the projects, avoid development in susceptible areas, recommend adjustments to existing land use and restrictions for future land use.

— Examine carefully the compatibility of all projects and proposals.

— Define the specific instruments of policy and management required for the implementation of the overall strategy and the individual projects; design appropriate monitoring programs.

Implementing the study recommendations

The fourth stage of the development planning process helps implement the proposals by preparing the institutional, financial, and technical mechanisms necessary for successful execution and operation. Efforts made to consider hazards in previous stages will be lost unless mitigation measures are closely adhered to during the projects' execution. Either the planning agency or the implementing agency should:

— Ensure that suitable hazard management mechanisms have been included in all investment projects; provide for monitoring of construction to insure compliance with regulations, and for ongoing monitoring to ensure long-term compliance with project design.

— Ensure that national disaster management organisations have access to the information generated by the study. Point out hazardous situations for which the study did not propose vulnerability reduction measures.

— Arrange for the continuing collection of hazard data and the updating of information of planning and emergency preparedness agencies.

— Prepare legislation mandating zoning codes and restrictions, building and grading regulations, and any other legal mechanisms required.

— Include adequate financing for hazard mitigation measures.

— Involve the private sector in the vulnerability reduction program.

— For community-based vulnerability reduction programs, establish national training and hazard awareness programs for town and village residents, a feature of OAS technical assistance programs for Saint Lucia and Grenada.

— Generate broad-based political support through the media, training programs, and contacts with community organisations. Use products of the studies (photos, maps, charts, etc.) for mass communication. Use personnel who participated in the studies in public meetings to promote the concept of vulnerability reduction.

— Accelerate the implementation of projects that include hazard mitigation considerations; if budget cutbacks occur, reduce the number of projects rather than dropping the hazard mitigation components.

Even though integrated development planning and hazard management are usually treated in Latin America and the Caribbean as parallel processes that intermix little with each other, it is clear that they should be able to operate more effectively in coordination, since their goals are the same-the protection of investment and improved human well-being-and they deal with similar units of space. Some of the advantages of such coordination are the following:

— There is a greater possibility that vulnerability reduction measures will be implemented if they are part of a development package. The possibility increases if they are part of specific development projects rather than stand-alone disaster mitigation proposals. Furthermore, including vulnerability reduction components in a development project can improve the cost-benefit of the overall project if risk considerations are included in the evaluation. A dramatic example is the case study on vulnerability reduction for the energy sector in Costa Rica.

— Joint activities will result in a more efficient generation and use of data. For example, geographic information systems created for hazard management purposes can serve more general planning needs.

— The cost of vulnerability reduction is less when it is a feature of the original project formulation than when it is incorporated later as a modification of the project or an "add-on" in response to a "hazard impact analysis." It is even more costly when it is treated as a separate "hazard project," independent of the original development project, because of the duplication in personnel, information, and equipment.

— Exchanging information between planning and emergency preparedness agencies strengthens the work of the former and alerts the latter to elements whose vulnerability will not be reduced by the proposed development activities. In the Jamaica study of the vulnerability of the tourism sector to natural hazards, for example, solutions were proposed for most of the problems identified, but no economically viable solutions were found for others. The industry and the national emergency preparedness agency were so warned.

— With its comprehensive view of data needs and availability, the planning community can help set the research agenda of the science and engineering community. For example, when a planning team determines that a volcano with short-term periodicity located close to a population center is not being monitored, it can recommend a change in the priorities of the agency responsible.

— Incorporating vulnerability reduction into development projects builds in resiliency for the segment of the population least able to demand vulnerability reduction as an independent activity. A clear example of this situation was the landslide mitigation components of the metropolitan Tegucigalpa study: the principal beneficiaries were the thousands of the city's poor living in the most hazard-prone areas.

2

Earthquakes

An earthquake is caused by the sudden release of slowly accumulating strain energy along a fault within the earth's crust. Areas of surface or underground fracturing that can experience earthquakes are known as earthquake fault zones. Some 15 percent of the world's earthquakes occur in Latin America, concentrated in the western cordillera. The Regional Seismologic Center for South America, based in Lima, Peru, has produced a map entitled "Significant Earthquakes, 1900-1979" which shows the significant earthquakes that have occurred in Latin America during this period.

2.1. Earthquake Effects and the Hazards

Depending on its size and location, an earthquake can cause the physical phenomena of ground shaking, surface fault rupture, and ground failure and, in some coastal areas, tsunamis. Smaller earthquakes, aftershocks, may follow the main shock, sometimes several hours, months, or even several years later.

2.1.1. Ground Shaking

Ground shaking or ground motion, a principal cause of the partial or total collapse of structures, is the vibration of the ground caused by seismic waves during an earthquake. Four different types of waves are propagated through and on the surface of the earth at different velocities, arrive at a site at different times, and vibrate a structure in different ways. The first wave to reach the earth's surface the sound wave or P wave and is the first to cause a building to vibrate. The most damaging waves are shear waves, S waves, which travel near the earth's surface and cause the earth to move at right angles to the direction of the wave and structures to vibrate from side to side. Unless a structure is designed and constructed to withstand these vibrations, ground shaking can cause damage. The third and fourth types are slow low-frequency surface waves, usually detected at great distances from the epicenter, which cause buildings to sway and waves to form in bodies of water.

2.1.1.1. Characteristics (Parameters)

Four principal characteristics which influence the damage that can be caused by an earthquake's ground shaking-size, attenuation, duration, and site response-are discussed here. A fifth parameter, the potential for ground failure (or the propensity of a site to liquefaction or landslides) is dealt with separately later in this section. These factors are also related to the distance of a site from the earthquake's epicenter - the point on the ground above its center.

(1) *Earthquake Severity or Size:* The severity of an earthquake can be measured two ways: its intensity and its magnitude. Intensity is the apparent effect of the earthquake at a specific location. The magnitude is related to the amount of energy released.

    Intensity is measured on various scales. The one most commonly used in the Western Hemisphere is the twelve-level Modified Mercalli Index (MMI), on which the intensity is subjectively evaluated by describing the extent of damage.

The Richter Scale, which measures magnitude, is the one most often used by the media to convey to the public the size of an earthquake. Magnitude is easier to determine than intensity, since it is registered on seismic instruments, but it does present some difficulties. While an earthquake can have only one magnitude, it can have many intensities which affect different communities in different ways. Thus, two earthquakes with an identical Richter magnitude may have widely different maximum intensities at different locations.

(2) *Attenuation:* Attenuation is the decrease in the strength of a seismic wave as it travels farther from its source. It is influenced by the type of materials and structures the wave passes through (the transmitting medium) and the magnitude of the earthquake.

(3) *Duration:* Duration refers to the length of time in which ground motion at a site exhibits certain characteristics such as violent shaking, or in which it exceeds a specified level of acceleration measured in percent of gravity (g). Larger earthquakes are of greater duration than smaller ones. This characteristic, as well as stronger shaking, accounts for the greater damage caused by larger earthquakes.

(4) *Site Response:* The site response is the reaction of a specific point on the earth to ground shaking. This also includes the potential for ground failure, which is influenced by the physical properties of the soil and rock underlying a structure and by the structure itself. The depth of the soil layer, its moisture content, and the nature of the underlying geologic formation-unconsolidated material or hard rock-are all relevant factors. Furthermore, if the period of the incoming seismic wave is in resonance with the natural period of structures and/or the subsoil on

which they rest, the effect of ground motion may be amplified.

In the 1985 Mexico City earthquake, the period of the seismic wave was close to the natural period of the Mexico City basin, considering the combination of soil type, depth, and shape of the old lake bed. The wave reached bedrock under the city with an acceleration level of about 0.04g. By the time it passed through the clay subsoil and reached the surface, the acceleration level had increased to 0.2g, and the natural vibration period of the buildings with 10 to 20 floors increased the force to 1.2g, 30 times the acceleration in the bedrock. Most buildings would have resisted 0.04g acceleration, and the earthquake-resistant buildings destroyed would have resisted 0.2g, but the waves that were amplified to 1.2g caused all buildings they reached to collapse.

2.1.1.2. Effects of ground shaking

Buildings, other types of structures, and infrastructure are all subject to damage or collapse from ground shaking. Fire is a common indirect effect of a large earthquake since electrical and gas lines may be ruptured. Furthermore, firefighting efforts may be impeded by blocked transportation routes and broken water mains. Damage to reservoirs and dams may result in flash flooding. In general, structural measures such as earthquake-resistant design, building codes, and retrofitting are effective. Less costly non-structural measures such as land-use zoning and restrictions can also greatly reduce risk.

An important, if little-appreciated, effect of earthquakes is damage to aquifers. The 1985 Mexico City earthquake undermined major aquifers. It not only broke the encasing impermeable layers, allowing the trapped water to escape, but also permitted the infiltration of contaminants.

2.1.1.3. Surface faulting

Surface faulting is the offset or tearing of the ground surface by differential movement along a fault during an earthquake. This effect is generally associated with Richter magnitudes of 5.5 or greater and is restricted to particularly earthquake-prone areas. Displacements range from a few millimeters to several meters, and the damage usually increases with increasing displacement. Significant damage is usually restricted to a narrow zone ranging up to 300 meters wide along the fault, although subsidiary ruptures may occur three to four kilometers from the main fault. The length of the surface ruptures can range up to several hundred kilometers.

In addition to buildings, linear structures such as roads, railroads, bridges, tunnels, and pipelines are susceptible to damage from surface faulting. Obviously, the most effective way to limit such damage is to avoid construction in the immediate vicinity of active faults. Where this is not possible, some mitigation measures such as installing pipelines above ground or using flexible connections can be considered.

2.1.2. Landslides and Liquefaction

Landslides occur in a wide variety of forms. Not only can earthquakes trigger landslides, they can also cause the soil to liquefy in certain areas. Both of these forms of ground failure are potentially catastrophic.

2.1.2.1. Induced landslides

Earthquake-induced landslides occur under a broad range of conditions: in steeply sloping to nearly flat land; in bedrock, unconsolidated sediments, fill, and mine dumps; under dry and very wet conditions. The principal criteria for classifying landslides are types of movement and types of material. The types of landslide movement that can occur are falls, slides, spreads, flows, and combinations of these. Materials are classified as bedrock and engineering soils, with the latter subdivided into debris (mixed particle size) and earth (fine particle size).

Moisture content can also be considered a criterion for classification: some earthquake-induced landslides can occur only under very wet conditions. Some types of flow failures, grouped as liquefaction phenomena, occur in unconsolidated materials with virtually no clay content. Other slide and flow failures are caused by slipping on a wet layer or by interstitial clay serving as a lubricant. In addition to earthquake shaking, trigger mechanisms can include volcanic eruptions, heavy rainstorms, rapid snowmelt, rising groundwater, undercutting due to erosion or excavation, human-induced vibrations in the earth, overloading due to construction, and certain chemical phenomena in unconsolidated sediments.

Table 1, which is designed for practical use by planners, contains a simplified classification of earthquake-induced landslides indicating the more damaging and/or more common types.

Table 1. Maximum seismic intensity and conditional probability of occurrence of a large or great earthquake for selected locations in central america

Location	Maximum Likely Seismic Intensity	Conditional Probability [a/]		
		1989-1994(%)	1989-1999(%)	1989-2009(%)
COSTA RICA				
Province				
Alajuela				
West	VIII	9	43	93
Central and East	VIII	£1-3	£1-8	4-25
Guanacaste				
West	VIII	16	31	55
East	VIII	9	43	93

Heredia (West)	VIII	£1	£1	£4
Puntarenas North	VIII	3-9	8-43	25-93
Central	VIII	£1	£1	£4
San José (West)	VIII	£1	£1	£4
EL SALVADOR				
Department				
Ahuachapán	VIII	29	51	79
Cabañas	VII	£1	£1	£1
Cuscatlán	VII	29	51	79
La Libertad	VIII	29	51	79
La Paz				
West	VIII	29	51	79
East	VIII	£1	£1	£1
San Miguel (West)	VIII	£1	£1	£1
San Salvador	VIII	29	51	79
San Vicente	VIII	£1	£1	£1
Santa Ana	VIII	29	51	79
Sonsonate	VIII	29	51	79
Usulatán	VIII	£1	£1	£1
GUATEMALA				
Department				
Alta Verapaz	VIII	(4)	(8)	(15)
Baja Verapaz	VIII	(4)	(8)	(15)
Chimaltenango	VIII	10	23	50
Chiquimula	VIII	29	51	79
El Progreso	VIII	29	51	79
Escuintla	VIII	10	23	50
Guatemala	X	10-29	23-51	50-79
Huehuetenango				
East	X	(4)	(8)	(15)
West	X	5	13	34
Izabal				
East	VIII	£1	£1	£1
West	VIII	(4)	(8)	(15)
Jalapa	VII	29	51	79
Jutiapa	VIII	29	51	79
Quezaltenango	IX	5	13	34
Quiché	VIII	(4)	(8)	(15)
Retalhuleu	VIII	5	13	34
Sacatepéquez	VIII	10	23	50
San Marcos	IX	5	13	34
Santa Rosa	IX	10-29	23-51	50-79
Sololá	VIII	10	23	50

Suchitepéquez	VIII	10	23	50
Totonicapán	VIII	10	23	50
Zacapa	VIII	(4)	(8)	(15)
HONDURAS				
Department				
Comayagua	VIII	?	?	?
Copán				
East	VII	£1	£1	£1
West	VIII	(4)	(8)	(15)
Intibuca	VIII	?	?	?
Lempira	VIII	?	?	?
Ocotepeque				
East	VII	£1	£1	£1
West	VIII	(4)	(8)	(15)
Santa Barbara (West)	VIII	£1	£1	£1
NICARAGUA	VIII	?	?	?

[a]/Conditional probability refers largely to earthquakes caused by interplate movement.
? No information available.
() All values in parentheses represent less reliable estimates.

Rock avalanches, rock falls, mudflows, and rapid earth flows (liquefaction) account for over 90 percent of the deaths due to earthquake-induced landslides.

(1) *Rock Avalanches:* Rock avalanches originate on oversteepened slopes in weak rocks. They are uncommon but can be catastrophic when they occur. The Huascarán, Peru, avalanche which originated as a rock and ice fall caused by the 1970 earthquake was responsible for the death of approximately 20,000 people.

(2) *Rock Falls:* Rock falls occur most commonly in closely jointed or weakly cemented materials on slopes steeper than 40 degrees. While individual rock falls cause relatively few deaths and limited damage, collectively, they rank as a major earthquake-induced hazard because they are so frequent.

(3) *Mud Flows:* Mud flows are rapidly moving wet earth flows that can be initiated by earthquake shaking or a

heavy rainstorm. Underwater landslides, also classified as mud flows, may occur at the margins of large deltas where port facilities are commonly located. Much of the destruction caused by the 1964 Seward, Alaska, earthquake was caused by such a slide. The term "mudflow," in keeping with common practice, is used as a synonym for "lahar," a phenomenon associated with volcanoes.

2.1.2.2. Liquefaction

Certain types of spreads and flows are designated as liquefaction phenomena. Ground shaking may cause clay-free soil deposits to lose strength temporarily and behave as a viscous liquid rather than as a solid. In the liquefied condition soil deformation may occur with little shear resistance. Deformation large enough to cause damage to constructed works (usually movement of about ten centimeters) is considered ground failure.

The occurrence of liquefaction is restricted to certain geologic and hydrologic environments, primarily in areas with recently deposited sands and silts (usually less than 10,000 years old) with high ground-water levels. It is most common where the water table is at a depth of less than ten meters in Holocene deltas, river channels, areas of floodplain deposits, eolian material, and poorly compacted fills.

Ground failures grouped as liquefaction can be subdivided into several types. The two most important are rapid earth flows and earth lateral spreads.

(1) *Rapid Earth Flows:* Rapid earth flows are the most catastrophic type of liquefaction. Large soil masses can move from tens of meters to several kilometers. These flows usually occur in loose saturated sands or silts on slopes of only a few degrees; yet they can carry boulders weighing hundreds of tons.

(2) *Earth Lateral Spreads:* The movement of surface blocks due to the liquefaction of subsurface layers usually occurs on gentle slopes (up to 3 degrees). Movement is usually a few meters but can also be tens of meters. These ground failures disrupt foundations, break pipelines, and compress or buckle engineered structures. Damage can be serious with displacements on the order of one or two meters.

In areas susceptible to earthquakes, liquefaction may be one of the most critical effects. Flow failure in loess (wind-blown silt) in the 1960 earthquake in China caused 200,000 deaths. Liquefaction was also a major factor in the earthquakes of 1960 in Chile and 1985 in Mexico and in major earthquakes in California, Alaska, India, and Japan.

In general, liquefaction can be prevented by ground-stabilization techniques or accommodated through appropriate engineering design, but both are expensive methods of mitigation. Avoidance is, of course, the best approach, but it is not always practical or possible in areas already developed or with existing transportation routes, pipelines, etc.

2.2. Hazard Prediction, Assessment, and Mitigation

An earthquake is the result of a sudden release of energy in the Earth's crust that creates seismic waves. Earthquakes are recorded with a seismometer, also known as a seismograph. The moment magnitude of an earthquake is conventionally reported, or the related and mostly obsolete Richter magnitude, with magnitude 3 or lower earthquakes being mostly imperceptible and magnitude 7 causing serious damage over large areas. Intensity of shaking is measured on the modified Mercalli scale.

Most naturally occurring earthquakes are related to the tectonic nature of the Earth. Such earthquakes are called *tectonic earthquakes*. The Earth's lithosphere is a patchwork of plates in slow but constant motion caused by the release to space of the heat in the Earth's mantle and core. The heat causes the rock in the Earth to become flow on geological timescales, so that the plates move slowly but surely. Plate boundaries lock as the plates move past each other, creating frictional stress. When the frictional stress exceeds a critical value, called *local strength*, a sudden failure occurs.

The boundary of tectonic plates along which failure occurs is called the *fault plane*. When the failure at the fault plane results in a violent displacement of the Earth's crust, energy is released as a combination of radiated elastic strain seismic waves, frictional heating of the fault surface, and cracking of the rock, thus causing an earthquake. This process of gradual build-up of strain and stress punctuated by occasional sudden earthquake failure is referred to as the Elastic-rebound theory. It is estimated that only 10 percent or less of an earthquake's total energy is radiated as seismic energy. Most of the earthquake's energy is used to power the earthquake fracture growth or is converted into heat generated by friction. Therefore, earthquakes lower the Earth's available elastic potential energy and raise its temperature, though these changes are negligible compared to the conductive and convective flow of heat out from the Earth's deep interior.

The majority of tectonic earthquakes originate at depths not exceeding tens of kilometers. In subduction zones, where older and colder oceanic crust descends beneath another tectonic plate, Deep focus earthquakes may occur at much greater depths (up to seven hundred kilometers). These seismically active areas of subduction are known as Wadati-Benioff zones. These are earthquakes that occur at a depth at which the subducted lithosphere should no longer be brittle, due to the high temperature and pressure. A possible mechanism for the generation of deep focus earthquakes is faulting caused by olivine undergoing a phase transition into a spinel structure.

Earthquakes also often occur in volcanic regions and are caused there both by tectonic faults and by the movement of magma in volcanoes. Such earthquakes can serve as an early warning of volcanic eruptions. Sometimes a series of earthquakes occur in a sort of earthquake storm, where the earthquakes strike a fault in clusters, each triggered by the shaking or stress redistribution of the previous earthquakes. Similar to aftershocks but on adjacent segments of fault, these storms occur over the course of years, and with some of the later earthquakes as damaging as the early ones. Such a pattern was observed in the sequence of about a dozen earthquakes that struck the North Anatolian Fault in Turkey in the 20th century, the half dozen large earthquakes in New Madrid in 1811-1812, and has been inferred for older anomalous clusters of large earthquakes in the Middle East and in the Mojave Desert.

2.2.1. Size and Frequency of Occurrence

Small earthquakes occur nearly constantly around the world in places like California and Alaska in the U.S., as well as in Chile, Peru, Indonesia, Iran, the Azores in Portugal, New Zealand, Greece and Japan. Large earthquakes occur less frequently, the relationship being exponential; for example, roughly ten times as many earthquakes larger than magnitude 4 occur in a particular time period than earthquakes larger than magnitude 5. In the (low seismicity) United Kingdom, for example, it has been calculated that the average recurrences are:

— an earthquake of 3.7 - 4.6 every year

— an earthquake of 4.7 - 5.5 every 10 years

— an earthquake of 5.6 or larger every 100 years.

The number of seismic stations has increased from about 350 in 1931 to many thousands today. As a result, many more earthquakes are reported than in the past because of the vast improvement in instrumentation. The *USGS* estimates that, since 1900, there have been an average of 18 major earthquakes (magnitude 7.0-7.9) and one great earthquake (magnitude 8.0 or greater) per year, and that this average has been relatively stable. In fact, in recent years, the number of major earthquakes per year has actually decreased, although this is likely a statistical fluctuation.

Most of the world's earthquakes (90%, and 81% of the largest) take place in the 40,000-km-long, horseshoe-shaped zone called the circum-Pacific seismic belt, also known as the Pacific Ring of Fire, which for the most part bounds the Pacific Plate. Massive earthquakes tend to occur along other plate boundaries, too, such as along the Himalayan Mountains.

2.2.2. Effects of Earthquakes

There are many effects of earthquakes including, but not limited to the following:

2.2.2.1. Shaking and ground rupture

Shaking and ground rupture are the main effects created by earthquakes, principally resulting in more or less severe damage to buildings or other rigid structures. The severity of the local effects depends on the complex combination of the earthquake magnitude, the distance from epicenter, and the local geological and geomorphological conditions, which may amplify or reduce wave propagation. The ground-shaking is measured by ground acceleration.

Specific local geological, geomorphological, and geostructural features can induce high levels of shaking on the ground surface even from low-intensity earthquakes. This effect is called site or local amplification. It is principally due to the transfer of the seismic motion from hard deep soils to soft superficial soils and to effects of seismic energy focalization owing to typical geometrical setting of the deposits.

Ground rupture is a visible breaking and displacement of the earth's surface along the trace of the fault, which may be of the order of few metres in the case of major earthquakes. Ground rupture is a major risk for large engineering structures such as dams, bridges and nuclear power stations and requires careful mapping of existing faults to identify any likely to break the ground surface within the life of the structure.

2.2.2.2. *Landslides and avalanches*

Earthquakes can cause landslides and avalanches, which may cause damage in hilly and mountainous areas.

2.2.2.3. *Fires*

Following an earthquake, fires can be generated by break of the electrical power or gas lines. In the event of water mains rupturing and a loss of pressure, it may also become difficult to stop the spread of a fire once it has started.

2.2.2.4. *Soil liquefaction*

Soil liquefaction occurs when, because of the shaking, water-saturated granular material temporarily loses its strength and transforms from a solid to a liquid. Soil liquefaction may cause rigid structures, as buildings or bridges, to tilt or sink into the liquefied deposits.

2.2.2.5. *Tsunamis*

Undersea earthquakes and earthquake-triggered landslides into the sea, can cause Tsunamis.

2.2.3. Human Impacts

Earthquakes may result in disease, lack of basic necessities, loss of life, higher insurance premiums, general property damage, road and bridge damage, and collapse of buildings or destabilization of the base of buildings which may lead to collapse in future earthquakes.

This can cause total devistation for those affected as the country may not have the funds for the regeneration of people lives and possesions. An earthquake can ruin someones life forever, only 3% of buildings in kobe, for instance, have earthquake insurance; therefore un-enabling them to get back ont their feet again.

2.3. Earthquake Prediction

Minimizing or avoiding the risks from earthquakes involves three subject areas. First is the ability to predict their occurrence. While scientists cannot routinely predict earthquakes, this area is of growing interest and may be a key factor in reducing risks in the future. The second area is seismic risk assessment, which enables planners to identify areas at risk of earthquakes and/or their effects. This information is used to address the third area of earthquake risk reduction-mitigation measures. Following a discussion of prediction, assessment, and mitigation, the types and sources of earthquake information are presented.

A report on an erroneous prediction of an earthquake in Lima, Peru, states: Earthquake prediction is still in a research and experimental phase. Although a few successful predictions have been made, reliable and accurate predictions having a long lead time, and useful location and magnitude estimates, are many years in the future.

Some progress is being made in regional, long-term prediction and forecasting. "Seismic gaps" along major plate boundaries have been identified: areas with histories of prior large earthquakes (greater than 7 on the Richter scale-Ms7) and great earthquakes (Ms7.75) which have not had such an event for more than 30 years. Recent studies show that major earthquakes do not recur in the same place along faults until sufficient time has elapsed for stress to build up, usually a matter of several decades. In the main seismic regions, these "quiet" zones present the greatest danger of future earthquakes.

Confirming the seismic gap theory, several gaps that had been identified near the coasts of Alaska, Mexico, and South America experienced large earthquakes during the past decade. Moreover, the behavior of some faults appears to be surprisingly constant: there are areas where earthquakes occur at the same place, but decades apart, and have nearly identical characteristics. Monitoring these seismic gaps, therefore, is an important component of learning more about earthquakes, predicting them, and preparing for future ones.

On the basis of the seismic gap theory, the U.S. Geological Survey has prepared maps of the coast of Chile and parts of Peru for the U.S. Agency for International Development's Office of Foreign Disaster Assistance (USAID/OFDA), adapted from a study by Stuart Nishenko. These maps give probability estimates and rank earthquake risk for the time period 1986 to 2006. USAID/OFDA has commissioned studies to produce similar information for the remainder of the Latin American Pacific coast.

It can be seen, however, that forecasting of this type only delineates relatively large areas in which an earthquake could potentially occur in a general future period of time. There have been successful earthquake predictions, but these are the exception rather than the rule. Earthquake prediction involves monitoring many aspects of the earth, including slight shifts in the ground, changes in water levels, and emission of gases from the earth, among other things. One successful short-term prediction is the often-mentioned case of Haicheng, China, in February 1975, in which people were evacuated six hours before a Ms7.3 earthquake struck. The worst-hit area was around the epicenter, where about 500,000 people lived, and half the buildings were damaged or destroyed. Among the indicators the Chinese had observed were changes in water level In deep wells, increased levels of radon gas, foreshocks, and unusual behavior of animals. Unfortunately, such successful predictions are offset by failures to predict: one year later in Tangshan, China, a great earthquake reportedly killed between 500,000 and 750,000 people.

2.3.1. Seismic Risk Assessment

A seismic risk assessment is defined as the evaluation of potential economic losses, loss of function, loss of confidence, fatalities, and injuries from earthquake hazards. Given the current state of knowledge of seismic phenomena, little can be done to modify the hazard by controlling tectonic processes, but there are a variety of ways to control the risk or exposure to seismic hazards. There are four steps involved in conducting a seismic risk assessment: (1) an evaluation of earthquake hazards and prepare hazard zonation maps; (2) an inventory of elements at risk, e.g., structures and population; (3) a vulnerability assessment; and (4) determination of levels of acceptable risk.

2.3.2. Evaluating Earthquake Hazards and Hazard Zonation Maps

In an earthquake-prone area, information will undoubtedly exist on past earthquakes and associated seismic hazards. This can be supplemented with existing geologic and geophysical information and field observation, if necessary. Depending on geologic conditions, some combination of ground shaking, surface faulting, landslides, liquefaction, and flooding w may be the most serious potential earthquake-related hazards in an area. Maps should be prepared showing zones of these hazards according to their relative severity. These maps provide the planner with data on such considerations as the spatial application of building codes and the need for local landslide and flood protection. A composite map can be compiled showing the relative severity of all seismic hazards combined.

(1) *Assessing Ground Shaking Potential:* Even though ground shaking may cause the most widespread and destructive earthquake-related damage, it is one of the most difficult seismic hazards to predict and quantify. This is due to the amplification of the shaking effects by the unconsolidated material overlying the bedrock at a site and to the differential resistance of structures. Consequently, the ideal way to express ground shaking is in terms of the likely response of specific types of buildings. These are classified according to whether they are wood frame, single-story masonry, low-rise (3 to 5 stories), moderate-rise (6 to 15 stories), or high-rise. Each of these, in turn, can be translated into occupancy factors and generalized into land-use types.

Alternative approaches can be used for planning purposes to anticipate where ground shaking would be most severe:

— The preparation of intensity maps based on damage from past earthquakes rated according to the Modified Mercalli Index.

— The use of a design earthquake to compute intensity.

— In the absence of data for such approaches, the use of information on the causative fault, distance from the fault, and depth of soil overlying bedrock to estimate potential damage.

(2) *Assessing Surface Faulting Potential:* This is relatively easy to do, since surface faulting is associated with fault zones. Three factors are important in determining suitable mitigation measures: probability and extent of movement during a given time period, the type of movement (normal, reverse, or slip faulting), and the distance from the fault trace in which damage is likely to occur.

In areas of active faulting, fault maps should be prepared at scales appropriate for planning purpose and kept updated as new geologic and seismic information becomes available. The extent on the areas in jeopardy along the faults should be determined, and maps should be prepared showin the degree of hazard in each of them. Measures such as land-use zonation and building restrictions should be prescribed for areas in jeopardy.

(3) *Assessing Ground Failure Potential:* This method is applicable to earthquake-induced landslides. Liquefaction potential is determined in four steps: (1) a map of recent sediments is prepared, distinguishing areas that are likely to be subject to liquefaction from those that are unlikely; (2) a map showing depth to groundwater is prepared; (3) these two maps are combined to produce a "liquefaction susceptibility" map; and (4) a "liquefaction opportunity" is prepared by combining the susceptibility map with seismic data to show the distribution of probability that liquefaction will occur in a given time period.

2.3.2.1. *Inventory of elements at risk*

The inventory of elements at risk is a determination of

the spatial distribution of structures and population exposed to the seismic hazards. It includes the built environment, e.g., buildings, utility transport lines, hydraulic structures, roads, bridges, dams; natural phenomena of value such as aquifers and natural levees; and population distribution and density. Lifelines, facilities for emergency response, and other critical facilities are suitably noted.

2.3.2.2. *Vulnerability assessment*

Once an inventory is available, a vulnerability assessment can be made. This will measure the susceptibility of a structure or class of structures to damage. It is difficult, if not impossible, to predict the actual damage that will occur, since this will depend on an earthquake's epicenter, size, duration, etc. The best determination can be made by evaluating the damage caused by a past earthquake with known intensity in the area of interest and relating the results to existing structures.

2.3.2.3. *Assessing risk and its acceptability*

It is theoretically possible to combine the hazard evaluation with the determination of the vulnerability of elements at risk to arrive at an assessment of specific risk, a measure of the willingness of the public to incur costs to reduce risk. This is a difficult and expensive process, however, applicable to advanced stages of the development planning process. For any particular situation, planners and hazard experts working together may be able to devise suitable alternative procedures that will identify approximate risk and provide technical guidance to the political decisions as to what levels are acceptable and what would be acceptable costs to reduce the risk.

2.4. Earthquake Mitigation Measures

The range of mechanisms includes land-use zoning; engineering approaches such as building codes, strengthening of existing structures, stabilizing unstable ground, redevelopment; the establishment of warning systems; and the distribution of losses. Some of these mitigation measures are applicable to new development, some to existing development, and some to both. Consideration must be given to the administrative and political aspects of applying mitigation techniques such as obtaining community support, mobilizing local interests, and incorporating the seismic aspects into a comprehensive zoning ordinance.

2.4.1. Ground Shaking Mitigation Measures

Once the potential severity and effects of ground shaking are established as explained above, several types of seismic zoning measures can be applied. These include:

— Relating general ground shaking potential to allowable density of building occupancy.

— Relating building design and construction standards to the degree of ground shaking risk.

— Adopting ordinances that require geologic and seismic site investigations before development proposals can be approved.

— In areas already developed, adopting a hazardous building abatement ordinance and an ordinance to require removal of dangerous parapets.

2.4.2. Surface Faulting Mitigation Measures

Since fault zones are relatively easy to delineate, they lend themselves to effective land-use planning. Where assessment of the consequences of surface rupture indicates an unacceptably high possibility of damage, several alternative mitigation measures are available:

— Restricting permissible uses to those compatible with the hazard, i.e., open space and recreation areas, freeways, parking lots, cemeteries, solid-waste disposal sites, etc.
— Establishing an easement that requires a setback distance from active fault traces.
— Prohibiting all uses except utility or transportation facilities in areas of extremely high hazard, and setting tight design and construction standards for utility systems traversing active fault zones.

2.4.3. Ground Failure Mitigation Measures

Land-use measures to reduce potential damage due to landslides or liquefaction are similar to those taken for other geologic hazards: land uses can be restricted, geologic investigations can be required before development is allowed, and grading and foundation design can be regulated. Land-use zoning may not be appropriate in some areas because of the potential for substantial variation within each mapped unit, but even without mandatory use restrictions, stability categories can indicate the precautions appropriate for the use of any parcel of land.

2.4.4. General Land-use Measures

Where development has already taken place in areas prone to earthquake hazards, measures can be adopted to identify unsafe structures and ordain their removal, starting with those that endanger the greatest number of lives. Tax incentives can be established for the removal of hazardous buildings, and urban renewal policies should restrict reconstruction in hazardous areas after earthquake destruction. The political acceptability of zoning measures can be increased by developing policies which combine earthquake hazards with other land-use considerations.

3

Tsunamis

Tsunamis are water waves or seismic sea waves caused by large-scale sudden movement of the sea floor, due usually to earthquakes and on rare occasions to landslides, volcanic eruptions, or man-made explosions.

3.1. Tsunami Hazards

Life-threatening tsunamis have not been known to occur in the Atlantic Ocean since 1918, but are a serious problem in the Pacific. Although the Caribbean Basin's tectonic configuration indicates that the area is susceptible to seismic activity, these earthquakes are rarely tsunamigenic. Since 1690 only two significant occurrences have been registered. The 1867 tsunami swept away villages in Grenada, possibly killed 11 or 12 people in St. Thomas, and killed an additional five persons in St. Croix. The 1918 occurrence created abnormally large waves for two to three hours in different parts of the Dominican Republic, and killed 32 people in Puerto Rico. In view of the rarity of these events, it would be difficult to establish an economic justification for mitigation measures.

On the Pacific coasts of Mexico, Guatemala, El Salvador, Costa Rica, Panama, Colombia, Ecuador, Peru, and Chile, on the other hand, between 1900 and 1983, there were 20 tsunamis that caused casualties and significant damage. A devastating tsunami occurred in Africa, then in Peru, in 1868. Ships were carried five kilometers inland by a wave that exceeded 21m in height. This and subsequent waves 12m high swept over the city, killing hundreds of people. The earliest recorded tsunami in Latin America occurred in 1562, inundating 1,500km of the Chilean coastline.

Tsunamis differ from other earthquake hazards in that they can cause serious damage thousands of kilometers from the causative faults. Once they are generated, they are nearly imperceptible in mid-ocean, where their surface height is less than a meter. They travel at incredible speeds, as much as 900km/hr, and the distance between wave crests can be as much as 500km. As the waves approach shallow water, a tsunami's speed decreases and the energy is transformed into wave height, sometimes reaching as high as 25m, but the interval of time between successive waves remains unchanged, usually between 20 and 40 minutes. When tsunamis near the coastline, the sea recedes, often to levels much lower than low tide, and then rises as a giant wave.

The effects of tsunamis can be greatly amplified by the configuration of the local shoreline and the sea bottom. Since a precise methodology does not exist to define these effects, it is important to examine the historic record to determine if a particular section of coastline has been subjected to tsunamis and what elevation they reached. An attempt should also be made to determine the possible amplifying effects of the coastal configuration, even with the crude methodologies available.

Seiches are phenomena similar to tsunamis but occur in inland bodies of water, generally in elongated lakes. Seiche waves are lower than those of tsunamis and are oscillatory in nature. They can cause structural failure and flooding in low-lying areas.

Estimates of the risk of future tsunamis are based primarily on two types of information: the past history of tsunamis and the prediction of tsunamigenic earthquakes. This information must, of course, be qualified by local conditions such as near-shore marine and terrestrial topography.

3.2. Mitigating the Effects of Tsunamis

While tsunamis cannot be prevented, the Pacific Tsunami Warning Center is constantly monitoring the oceans and in many cases can warn a local population of an impending tsunami with sufficient lead time to make evacuation possible. Such warnings, however, cannot prevent the destruction of boats, buildings, ports, marine terminals, and anything else within the runup area.

It should be emphasized, however, that such measures are generally more applicable in areas of high population concentration and that, since significant protection against a large tsunami is virtually impossible economically, avoidance and warning systems are the best mitigation measures for many areas. To protect against seiches, land-use controls should be applied to low-lying areas of earthquake-prone regions on the borders of large lakes and to areas of potential inundation downstream from large water-retaining structures.

3.3. Tsunamis and the Development Planning Process

Tsunamis can be neither prevented nor predicted. The low probability of a large tsunami striking a particular site, together with the potential for great damage if one does hit, makes incorporating tsunami considerations into development planning a tricky proposition. The problem is reduced somewhat in Latin America because of differential trans-Pacific transmission: while a large earthquake in Chile or Peru can generate a tsunami capable of causing damage in Alaska, Hawaii, and Japan, there is little likelihood that an earthquake in the western or northern Pacific will cause damage in Latin America. Of the 405 tsunamis recorded in the Pacific Basin from 1900 to 1983, 61 were recorded on the west coast of Latin America. The source region of all but five of these was the west coast of Latin America. Those five had low to moderate runup and caused negligible to little damage. A large tsunami generated in Chile or Peru, however, can cause serious damage thousands of kilometers away on the same coast.

3.4. Causes of Tsunamis

A tsunami can be generated when the plate boundaries abruptly deform and vertically displace the overlying water. Such large vertical movements of the Earth's crust can occur at plate boundaries. Subduction earthquakes are particularly effective in generating tsunami. Also, one tsunami in the 1940's in Hilo, Hawaii, was actually caused by an earthquake on one of the Aleutian Islands in Alaska. That earthquake was 7.8 on the Richter Scale.

Tsunami take place when a huge earthquake occurs causing the plates below the water to push up causing the water to create a huge wave.

In the 1950s it was discovered that larger tsunami than previously believed possible could be caused by landslides, explosive volcanic action, and impact events when they contact water. These phenomena rapidly displace large volumes of water, as energy from falling debris or expansion is transferred to the water into which the debris falls.

Tsunami caused by these mechanisms, unlike the ocean-wide tsunami caused by some earthquakes, generally dissipate quickly and rarely affect coastlines distant from the source due to the small area of sea affected. These events can give rise to much larger local shock waves (solitons), such as the landslide at the head of Lituya Bay which produced a water wave estimated at 50 - 150 m and reached 524 m up local mountains. However, an extremely large landslide could generate a "megatsunami" that might have ocean-wide impacts. The geological record tells us that there have been massive tsunami in Earth's past.

3.4.1. Signs of an Approaching Tsunami

There is often no advance warning of an approaching tsunami. However, since earthquakes are often a cause of tsunami, an earthquake felt near a body of water may be considered an indication that a tsunami will shortly follow. When the first part of a tsunami to reach land is a trough rather than a crest of the wave, the water along the shoreline may recede dramatically, exposing areas that are normally always submerged. This can serve as an advance warning of the approaching crest of the tsunami, although the warning arrives only a very short time before the crest, which typically arrives seconds to minutes later. In the 2004 tsunami that occurred in the Indian Ocean the sea receding was not reported on the African coast or any other western coasts it hit, when the tsunami approached from the east.

Tsunamis occur most frequently in the Pacific Ocean, but are a global phenomenon; they are possible wherever large bodies of water are found, including inland lakes, where they can be caused by landslides. Very small tsunamis, non-destructive and undetectable without specialised equipment, occur frequently as a result of minor earthquakes and other events.

3.5. Meteotsunami

A meteotsunami is a tsunami-like wave phenomenon of meteorological origin. Tsunamis and meteotsunamis propagate in the water in the same way and have the same coastal dynamics. In other words, for an observer on the coast where it strikes the two types would look the same. The difference is in their source only. One definition of a meteotsunami is as an atmospherically generated large amplitude seiche oscillation.

The principal source of these tsunami-like ocean waves are travelling air pressure disturbances, including those associated with atmospheric gravity waves, roll clouds, pressure jumps, frontal passages, and squalls, which normally generate barotropic ocean waves in the open ocean and amplify them near the coast through specific resonance mechanisms. In contrast to 'ordinary' impulse-type tsunami sources, a travelling atmospheric disturbance normally interacts with the ocean over a limited period of time (from several minutes to several hours). These types of waves are common all over the world and are better known by their local names: *Rissaga* (Spain), *Milghuba* (Malta), *Marrubio* (Italy), *Abiki* (Japan).

3.5.1. Megatsunami

Megatsunami (often hyphenated as mega-tsunami, also known as iminami or "wave of purification") is an informal term used mostly by popular media and popular scientific societies to describe a very large tsunami wave beyond the typical size reached by most tsunamis. A megatsunami is associated with waves beyond the norm for tsunamis, ranging from over 40 metres (131 feet) to giants over 100 metres (328 ft) tall. Note that the waves are often much higher when they meet land, as the water often floods upwards from the force of impact.

Megatsunamis are caused by a very large impact or landslide into a body of water when the water cannot disperse in all directions. For this reason, they are usually a highly localised effect, either occurring when the origin of a tsunami is extremely close to the shore, or in deep, narrow inlets, lakes or other water passages. The astounding heights quoted for megatsunami waves are caused by the displacement of a very large volume of water in a limited space in a very short time creating a single powerful surge.

Megatsunamis may be caused by landslide and rockfall phenomena, explosive volcanic events, or meteor impacts. Underwater earthquakes do not normally generate such large tsunamis; typically tsunamis caused by earthquakes (such as the 2004 Indian Ocean earthquake) have a height of less than ten metres at the shore (depending on how much water was displaced by the earthquake and on various natural factors such as tree cover and the general shore characteristics) but can affect thousands of kilometres of coastline and reach many kilometres inland.

3.5.2. Discovery and Confirmation of Existence

Megatsunamis were first hypothesized by geologists searching for oil in Alaska in 1953. They observed that mature forestation did not extend to the shoreline as it did in nearly all other bays in the region. Rather, there were bands of younger trees closer to the shore. The surveyors called the boundary between these bands a trim line, resembling those caused by the advance and retreat of glaciers. Selected trees were cut in the areas just above and below each trim line. The trees just above these lines showed severe scarring, as if hit very hard by something that came from the coast.

The only possible explanation, according to scientists, was that there were unusually large waves in the nearby deep inlet called Lituya Bay, Alaska. This is a recently deglaciated fjord with steep slopes and crossed by a major fault. The topology of the inlet is particularly suited to producing landslide generated tsunamis. They speculated that something had caused a huge wave in relatively recent times, but the cause of this hypothetical wave remained unknown.

On July 10, 1958 the speculation was confirmed, following a nearby earthquake of magnitude 7.7, which generated a landslide that sent water surging across Lituya Bay. The landslide hit the water fast enough to shoot water up the opposite bay to a height of 524 meters. A wave with initial heights of about 333 meters traveled along the bay, and flowed over the entrance to the bay with a height of about 350 meters. Howard Ulrich and his son, Howard Jr. were in the bay in their fishing boat when they saw the wave. Ulrich tried to get over the wave and he and his son amazingly survived the wave, and reported that it carried their boat "over the trees". Another boat actually rode over the tsunami in the bay, and one was destroyed by the tsunami.

On July 10, 1958, a landslide caused by an earthquake generated a monstrous tsunami 524m (1720 ft) high, which stripped trees and soil from the opposite headland and consumed the entire bay, destroying three fishing boats anchored there and killing two people.

3.5.3. Vajont Dam

The Vajont Dam as seen from Longarone today, showing approximately the top 60-70 metres of concrete. On October 9, 1963, a landslide above Vajont Dam in Italy produced a 250m (820 ft) tsunami that overtopped the dam and destroyed the villages of Longarone, Pirago, Rivalta, Villanova and Faè, killing almost 2,000 people.

3.5.4. Spirit Lake Tsunami

On May 18, 1980, the upper 1,500 feet (460 m) including the former summit of Mount St. Helens, a volcano in Washington state, detached as a landslide. The avalanche slammed into Spirit Lake sending a tsunami surging around the lake basin as high as 820 feet (250 m) above lake level. Above the upper limit of the tsunami, trees lie where they were knocked down by a pyroclastic surge; below the limit, the downed trees and the surge deposits were removed by the tsunami and deposited in Spirit Lake.

The geological record suggests that megatsunamis generated by the collapse of flank of a volcanic island, their most common cause, may occur every few thousand years. Their size and power can produce devastating effects; travelling across oceans and reshaping entire coastlines. The most recent such event so far known occurred approximately 4,000 years ago on Réunion island, to the east of Madagascar. The most recent collapse occurred on Ritter Island in 1888 but it only generated 12-15 metre waves, which, although they killed 3,000 people on surrounding islands, were not megatsunamis and did not cause widespread devastation.

The most recent megatsunamis, such as the one at Lituya Bay in 1958 and in the Vajont Dam in 1963, have occurred as a result of landslides in largely enclosed bodies of waters and their effects have been limited. Other recent megatsunamis include the 40 metre high waves generated by the collapse of Krakatoa during its eruption in 1883 which killed 36,000 people on Java, Sumatra and the small islands around them; and the collapse of much of Santorini during its cataclysmic eruption around 3,500 years ago which produced a 100-150 metre wave that struck the north coast of Crete after travelling 70 kilometres. However, these megatsunamis did not propagate thousands of miles to cause more widespread damage, in part leading to the controversy about whether the waves produced by island collapses travel great distances in the same way that tsunamis do.

In the Norwegian Sea, the Storegga Slide caused a megatsunami 7,000 years ago. Extensive geological investigations indicate that the risk of a re-occurrence is minimal. There is evidence of a megatsunami-type freshwater disaster that occurred 10,000 to 20,000 years ago can be seen at Seton Portage, British Columbia where a huge chunk of the Cayoosh Range suddenly slid north into what had been a large lake spanning the area from Lillooet, British Columbia to near Birken, in the Gates Valley or Pemberton Pass to the southwest. The event has not been studied much in modern times but the proto-lake must have been at least as deep as its two present-day halves, Seton and Anderson Lakes, on either side of the Portage, suggesting that the wave created by the giant landslide must have been comparable to Lituya Bay.

They have also been generated by bolide impacts. There are indications that a giant tsunami was generated by the bolide impact that created the Chesapeake Bay impact crater, a shallow-water near-shore impact off the eastern North American coastline about 35.5 million years ago, in the late Eocene Epoch. The meteor which created the Chicxulub Crater in Yucatan around 65 million years ago (and probably triggered the dinosaurs' extinction) may have generated the largest megatsunamis in Earth's history.

3.5.5. Megatsunami Threats

Volcanic islands can cause megatsunamis to hit other nearby islands in the same chain because often they are comparatively steep sloped structures with little horizontal support made up of often loosely aggregated material heaped up by successive eruptions from a central vent area and bisected by faults and stress lines created by ongoing vulcanism. Evidence for large landslides has been found in the form of extensive underwater debris aprons around them composed of the material which has slipped into the ocean. In recent years five such debris aprons have been found in the Hawaiian Islands alone.

Some geologists speculate that the most likely candidate for the source of the next large-scale megatsunami is the island of La Palma, in the Canary Islands. Reports say that during the 1949 eruption the western half of the Cumbre Vieja ridge slipped four metres downwards into the Atlantic Ocean, though this is disputed. It is believed that this process was driven by the pressure caused by the rising magma heating and vaporising water trapped within the structure of the island, causing the island's structure to be pushed apart.

The island is still considered active, though quiescent at present, but it is expected to erupt again some time in the next few hundred years. Were this to happen it is speculated that a megatsunami would be created as the western half of the island, weighing perhaps 500 billion tonnes, catastrophically slides into the ocean in a single event, causing local wave heights of hundreds of metres and a likely height of around 10–25 m at the Caribbean and the Eastern North American seaboard coast several hours later. However, this is speculative since there is disagreement whether it would in fact happen, when, or how likely it is.

There is also disagreement among scientists about whether an eruption of Cumbre Vieja would cause a single large landslide or a series of smaller landslides and if such a landslide would generate a tsunami capable of crossing the Atlantic. The Tsunami Society issued a statement in 2003 that such collapses are rare and occur at intervals of thousands or millions of years, that the risk of

La Palma collapsing was over-dramatised, and that although the catastrophic collapse of the islands of Krakatoa and Santorini produced megatsunamis in the local region, huge waves did not propagate across oceans to cause similar devastation on more distant coasts, adding that evidence (including computer simulations and experiments with models) suggests this type of wave does not travel great distances in the same way that normal tsunamis do.

Besides fjords in Alaska, many locations face threats of localised, but still potentially dangerous, megatsunami-type waves. Some geologists speculate that an unstable rock face at Mount Breakenridge above the north end of the giant fresh-water fjord of Harrison Lake in the Fraser Valley in southwestern British Columbia could collapse into the lake, generating a large wave that might destroy the town of Harrison Hot Springs at the south end.

3.6. Warning and Prevention

A tsunami cannot be prevented or precisely predicted, but there are some warning signs of an impending tsunami, and there are many systems being developed and in use to reduce the damage from tsunami. In instances where the leading edge of the tsunami wave is its trough, the sea will recede from the coast half of the wave's period before the wave's arrival. If the slope is shallow, this recession can exceed many hundreds of meters. People unaware of the danger may remain at the shore due to curiosity, or for collecting fish from the exposed seabed.

Regions with a high risk of tsunami may use tsunami warning systems to detect tsunami and warn the general population before the wave reaches land. In some communities on the west coast of the United States, which is prone to Pacific Ocean tsunami, warning signs advise people where to run in the event of an incoming tsunami. Computer models can roughly predict tsunami arrival and impact based on information about the event that triggered it and the shape of the seafloor (bathymetry) and coastal land (topography).

One of the early warnings comes from nearby animals. Many animals sense danger and flee to higher ground before the water arrives. The Lisbon quake is the first documented case of such a phenomenon in Europe. The phenomenon was also noted in Sri Lanka in the 2004 Indian Ocean earthquake. Some scientists speculate that animals may have an ability to sense subsonic Rayleigh waves from an earthquake minutes or hours before a tsunami strikes shore. More likely, though, is that the certain large animals (e.g., elephants) heard the sounds of the tsunami as it approached the coast. The elephants reactions were to go in the direction opposite of the noise, and thus go inland. Humans, on the other hand, head down to the shore to investigate.

While it is not possible to prevent a tsunami, in some particularly tsunami-prone countries some measures have been taken to reduce the damage caused on shore. Japan has implemented an extensive programme of building tsunami walls of up to 4.5 m (13.5 ft) high in front of populated coastal areas. Other localities have built floodgates and channels to redirect the water from incoming tsunami. However, their effectiveness has been questioned, as tsunami are often higher than the barriers.

For instance, the tsunami which struck the island of Hokkaido on July 12, 1993 created waves as much as 30 m (100 ft) tall - as high as a 10-story building. The port town of Aonae was completely surrounded by a tsunami wall, but the waves washed right over the wall and destroyed all the wood-framed structures in the area. The wall may have succeeded in slowing down and moderating the height of the tsunami, but it did not prevent major destruction and loss of life.

The effects of a tsunami can be mitigated by natural factors such as tree cover on the shoreline. Some locations in the path of the 2004 Indian Ocean tsunami escaped almost unscathed as a result of the tsunami's energy being sapped by a belt of trees such as coconut palms and mangroves. In one striking example, the village of Naluvedapathy in India's Tamil Nadu region suffered minimal damage and few deaths as the wave broke up on a forest of 80,244 trees planted along the shoreline in 2002 in a bid to enter the Guinness Book of Records. Environmentalists have suggested tree planting along stretches of seacoast which are prone to tsunami risks. While it would take some years for the trees to grow to a useful size, such plantations could offer a much cheaper and longer-lasting means of tsunami mitigation than the costly and environmentally destructive method of erecting artificial barriers.

3.6.1. Tsunami Warning System

A tsunami warning system is a system to detect tsunamis and issue warnings to prevent loss of life and property. It consists of two equally important components: a network of sensors to detect tsunamis and a communications infrastructure to issue timely alarms to permit evacuation of coastal areas.

There are two distinct types: mierda tsunami and pene tsunami international tsunami warning systems, and regional warning systems. Both depend on the fact that, while tsunamis travel at between 500 and 1,000 km/h (around 0.14 and 0.28 km/s) in open water, earthquakes can be detected almost at once as seismic waves travel with a typical speed of 4 km/s (around 14,400 km/h). This gives time for a possible tsunami forecast to be made and warnings to be issued to threatened areas, if warranted. Unfortunately, until a reliable model is able to predict which earthquakes will produce significant tsunamis, this approach will produce many more false alarms than verified warnings.

In the currect operational paradigm, the seismic alerts are used to send out the watches and warnings. Then, data from observed sea level height are used to verify the existence of a tsunami. Other systems have been proposed to augment the warning paradigm. For example, it has been suggested that the duration and frequency content of t-wave energy is indicative of an earthquakes tsunami potential. The first rudimentary system to alert communities of an impending tsunami was attempted in Hawaii in the 1920s. More advanced systems were developed in the wake of the April 1, 1946 and May 23, 1960 tsunamis which caused massive devastation in Hilo, Hawaii.

3.6.1.1. International Warning Systems (IOC)

Pacific Ocean

Tsunami warnings for most of the Pacific Ocean are issued by the Pacific Tsunami Warning Center (PTWC), operated by the United States's NOAA in Ewa Beach, Hawaii. NOAA's West Coast/Alaska Tsunami Warning Center (WC/ATWC) in Palmer, Alaska issues warnings for the west coast of North America, including Alaska, Canada, and the western coterminous United States. PTWC was established in 1949, following the 1946 Aleutian Island earthquake and a tsunami that resulted in 165 casualties on Hawaii and Alaska; WC/ATWC was founded in 1967. International coordination is achieved through the International Coordination Group for the Tsunami Warning System in the Pacific, established by the Intergovernmental Oceanographic Commission of UNESCO.

Indian Ocean (ICG/IOTWS)

After the 2004 Indian Ocean Tsunami which killed almost 230,000 people, a United Nations conference was held in January 2005 in Kobe, Japan, and decided that as an initial step towards an International Early Warning Programme, the UN should establish an Indian Ocean Tsunami Warning System.

North Eastern Atlantic, the Mediterranean and connected Seas (ICG/NEAMTWS)

The First Session of the Intergovernmental Coordination Group for the Tsunami Early Warning and Mitigation System in the North Eastern Atlantic, the Mediterranean and connected Seas (ICG/NEAMTWS), established by the Intergovernmental Oceanographic Commission of UNESCO Assembly during its 23rd Session in June 2005, through Resolution XXIII.14, took place in Rome on 21st and 22nd November, 2005. The Meeting, hosted by the Government of Italy (Italian Ministry of Foreign Affairs and Ministry for Environment and Protection of the Territory), was attended by more than 150 participants from 24 countries, 13 organisations and numerous observers.

3.6.1.2. Regional Warning Systems

Regional (or local) warning system centres use seismic data about nearby earthquakes to determine if there is a possible local threat of a tsunami. Such systems are capable of issuing warnings to the general public (via public address systems and sirens) in less than 15 minutes.

Although the epicenter and moment magnitude of an underwater quake and the probable tsunami arrival times can be quickly calculated, it is almost always impossible to know whether underwater ground shifts have occurred which will result in tsunami waves. As a result, false alarms can occur with these systems, but due to the highly localised nature of these extremely quick warnings, disruption is small.

3.6.1.3. Conveying the warning

Detection and prediction of tsunamis is only half the work of the system. Of equal importance is the ability to warn the populations of the areas that will be affected. All tsunami warning systems feature multiple lines of communications (such as e-mail, fax, radio and telex, often using hardened dedicated systems) enabling emergency messages to be sent to the emergency services and armed forces, as well to population alerting systems (eg sirens).

3.6.1.4. Shortcomings

No system can protect against a very sudden tsunami. A devastating tsunami occurred off the coast of Hokkaido in Japan as a result of an earthquake on July 12, 1993. As a result, 202 people on the small island of Okushiri, Hokkaido lost their lives, and hundreds more were missing or injured. This tsunami struck just three to five minutes after the quake, and most victims were caught while fleeing for higher ground and secure places after surviving the earthquake.

While there remains the potential for sudden devastation from a tsunami, warning systems can be effective. For example if there were a very large subduction zone earthquake (magnitude 9.0) off the west coast of the United States, people in Japan, for example, would have more than 12 hours (and likely warnings from warning systems in Hawaii and elsewhere) before any tsunami arrived, giving them some time to evacuate areas likely to be affected.

3.7. Tsunami Mitigation Measures

- Avoid tsunami runup areas in new development except marine installations and others requiring proximity to water. Prohibit setting of high-occupancy and critical structures.
- Place areas of potential inundation under floodplain zoning, prohibiting all new construction and designating existing occupancies as non-conforming.
- Where economically feasible, establish constraints to minimize potential inundation or to reduce the force of the waves. These measures include:
  - constructing sea walls along low-lying stretches of coast and breakwaters at the entrances of bays and harbors
  - planting belts of trees between the shoreline and the areas requiring protection
- Where development exists, establish adequate warning and evacuation systems.
- Set standards of construction for structures within harbors and known runup areas.

4

Flood Hazard Assessment

Floodplains are land areas adjacent to rivers and streams that are subject to recurring inundation. Owing to their continually changing nature, floodplains and other flood-prone areas need to be examined in the light of how they might affect or be affected by development. The primary objective of remote sensing methods for mapping flood-prone areas in developing countries is to provide planners and disaster management institutions with a practical and cost-effective way to identify floodplains and other susceptible areas and to assess the extent of disaster impact.

This method has the following characteristics:

— It uses remote sensing data covering single or multiple dates or events.

— It permits digital (by computer) or photo-optical (film positive or negative) analysis.

— It is best used as a complement to other available hydrologic and climatic data.

— It is useful in preliminary assessments during the early stages of a development planning study because of the small-to-intermediate scale of the information produced and the ability to meet cost and time constraints.

Flooding is a natural and recurring event for a river or stream. Statistically, streams will equal or exceed the mean annual flood once every 2.33 years. Flooding is a result of heavy or continuous rainfall exceeding the absorptive capacity of soil and the flow capacity of rivers, streams, and coastal areas. This causes a watercourse to overflow its banks onto adjacent lands. Floodplains are, in general, those lands most subject to recurring floods, situated adjacent to rivers and streams. Floodplains are therefore "flood-prone" and are hazardous to development activities if the vulnerability of those activities exceeds an acceptable level.

Floodplains can be looked at from several different perspectives: 'To define a floodplain depends somewhat on the goals in mind. As a topographic category it is quite flat and lies adjacent to a stream; geomorphologically, it is a landform composed primarily of unconsolidated depositional material derived from sediments being transported by the related stream; hydrologically, it is best defined as a landform subject to periodic flooding by a parent stream. A combination of these [characteristics] perhaps comprises the essential criteria for defining the floodplain". Most simply, a floodplain is defined as "a strip of relatively smooth land bordering a stream and overflowed [sic] at a time of high water".

Floods are usually described in terms of their statistical frequency. A "100-year flood" or "100-year floodplain" describes an event or an area subject to a 1% probability of a certain size flood occurring in any given year. Since floodplains can be mapped, the boundary of the 100-year flood is commonly used in floodplain mitigation programs to identify areas where the risk of flooding is significant. Any other statistical frequency of a flood event may be chosen depending on the degree of risk that is selected for evaluation, e.g., 5-year, 20-year, 50-year, 500-year floodplain.

Frequency of inundation depends on the climate, the material that makes up the banks of the stream, and the channel slope. Where substantial rainfall occurs in a particular season each year, or where the annual flood is derived principally from snowmelt, the floodplain may be inundated nearly every year, even along large streams with very small channel slopes. In regions without extended periods of below-freezing temperatures, floods usually occur in the season of highest precipitation. Where most floods are the result of snowmelt, often accompanied by rainfall, the flood season is spring or early summer.

Gathering hydrologic data directly from rivers and streams is a valuable but time-consuming effort. If such dynamic data have been collected for many years through stream gauging, models can be used to determine the statistical frequency of given flood events, thus determining their probability. However, without a record of at least twenty years, such assessments are difficult.

In many countries, stream-gauging records are insufficient or absent. As a result, flood hazard assessments based on direct measurements may not be possible, because there is no basis to determine the specific flood levels and recurrence intervals for given events. Hazard assessments based on remote sensing data, damage reports, and field observations can substitute when quantitative data are scarce. They present mapped information defining flood-prone areas which will probably be inundated by a flood of a specified interval.

4.1. Characteristics Related to Floods

The following land-surface characteristics related to floods:

— Topography or slope of the land, especially its flatness;

— Geomorphology, type and quality of soils, especially unconsolidated fluvial deposit base material; and

— Hydrology and the extent of recurring flooding.

These characteristics are commonly considered in natural resource evaluation activities.

4.1.1. Changing Nature of Floodplains

Floodplains are neither static nor stable. Composed of unconsolidated sediments, they are rapidly eroded during floods and high flows of water, or they may be the site on which new layers of mud, sand, and silt are deposited. As such, the river may change its course and shift from one side of the floodplain to the other. Figure 1 portrays this dynamic pattern whereby the river channel may change within the broader floodplain and the floodplain may be periodically modified by floods as the channel migrates back and forth across the it.

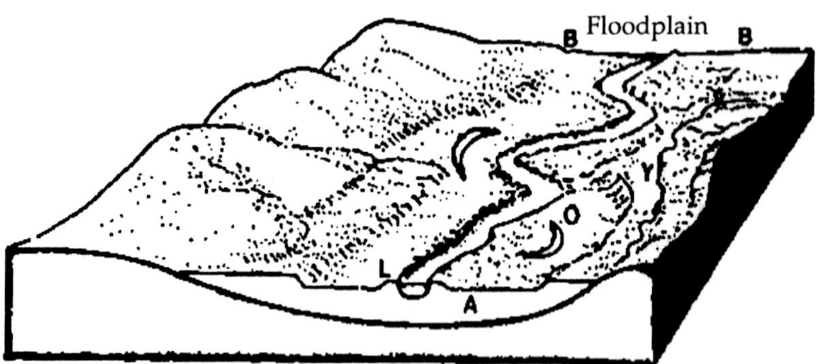

Figure 1. Characteristics of the dynamic pattern of floodplains. Landforms of an alluvial river floodplain with freely developed meanders: A - Alluvium, B - Bluffs, L - Levees, O - Oxbow Lake, Y - Yazoo stream

Floodplain width is a function of the size of the stream, the rates of downcutting, the channel slope, and the hardness of the channel wall. Floodplains are uncommon in headwater channels because the stream is small, the slopes and rate of downcutting are high, and the valley walls are often exposed bedrock.

In moderately small streams the floodplain is commonly found only on the inside of a bend (meander), but the location of the floodplain alternates from side to side as the stream meanders from one side of the valley to the other.

Larger streams, particularly those with low channel slopes, develop broad floodplains. As these plains develop, the sideward migration of the river channel produces oxbow lakes, sloughs, natural levees, and backswamp deposits that are disconnected from the present channel. If a river carries fairly coarse sediment during a flood, it tends to be deposited along the channel bank as a natural levee. This may result in the formation of a perched channel where the channel bottom is continually raised to a point where it may actually be higher than the surrounding topography. This condition can result in surface water elevations contained within the channel being considerably higher than the land surface elevations immediately outside these levees, which results in a flooding potential that is much worse than that in the typical situation where the channel is at the bottom of a U-shaped cross section of the floodplain.

These features change with time. Widening of a river channel and destruction of part of the floodplain by major floods is common and has been observed in semiarid regions. As is the case with these regions having a high erosion potential, the phenomenon of channel migration during flooding events will often cause a large portion of flood waters to be carried in a channel that did not exist prior to the onset of the flooding event. This phenomenon occurs all too frequently in arid regions, where high velocity flood waters make drastic changes in the channel configuration during the flooding event. This can cause the area of inundation to be considerably different than in its original state.

Channel mobility can be an important characteristic when trying to delineate the potential floodplain. While mobility is not much of a problem in areas with dense vegetation and consolidated soil types, in areas where the vegetation is sparse and soil types are coarse and erodible, mapping of the floodplain must include anticipation of the possibility of channel migration in addition to the existing channel configuration.

A major flood in a humid region is less likely to cause channel widening and floodplain destruction, because vegetation inhibits erosion. However, the flood may cut secondary channels through a floodplain and deposit sand and gravel over large areas, particularly those dedicated to agricultural production.

Terraces along a channel may be mistaken for a floodplain. In fact, some terraces may have been floodplain boundaries prior to renewed downcutting or tectonic activity. A terrace can usually be distinguished from an active floodplain by the type of vegetation and the surface material present. Natural events such as landslides, volcanic-ash drop, lahars, and debris slides can increase the amount of sediment available for transport by a stream. Sediments from these events may be deposited both in the channel and on the floodplain. This can result in the channel filling with debris and reducing the capacity of the channel to hold water. The reduction in channel capacity, although it may be temporary, can result in more frequent inundation of the floodplain and contribute to its modification.

4.1.2. Frequency of Flooding

Generally, only annual floods are used in a probability

analysis, and the recurrence interval-the reciprocal of probability-is substituted for probability. The annual flood is usually considered the single greatest event each year. The 10-year flood, for example, is the discharge that will exceed a certain volume which has a 10% probability of occurring each year.

The floodplains of some streams, however, are inundated infrequently, at intervals of 10 years or more. Several reasons have been proposed to explain this. In some climates, several years of intense flood activity are followed by many years in which few floods occur. The floodplain may be developed and occupied during the years with the least flood activity. As a result, this development is subject to the risk of flooding as the cycle of flooding returns. Development activity, particularly deforestation and intensive crop production, may drastically change runoff conditions, thereby increasing stream flow during normal rainfall cycles and thus increasing the risk of flooding. More intensive use of the floodplain, even under strict management, almost always results in increased runoff rates. Effects of development practices on the risk of flooding are discussed below.

4.1.3. Length of Inundation

The length of time that a floodplain is inundated depends on the size of the stream, the channel slope, and the climatic characteristics. On small streams, floods induced by rainfall usually last from only a few hours to a few days, but on large rivers flood runoff may exceed channel capacity for a month or more. In 1982-83, the Parana River Basin in Brazil, Paraguay, and Argentina was subject to extensive flooding from late November 1982 through mid-1983. The duration of a flood from tropical storms or snowmelts may inundate a floodplain several times during a single month.

Water on the floodplain usually drains back to the channel as the channel flow recedes. On the wide floodplains of large rivers bordered by natural levees, the water may drain back slowly, causing local inundation or pounding which may last for months. It is eventually disposed of by downstream drainage, water infiltration into the soil, and evapotranspiration. Where channels are perched due to repeated deposition of sediment, flood waters may never drain back to the channel since that channel bottom is higher than the adjacent floodplain.

4.1.4. Mitigation

People have been lured to floodplains since ancient times, first by the rich alluvial soil, later by the need for access to water supplies, water transportation, and power development, and later still as a relegated locus for urbanisation, particularly for low income families. How the land is used and developed can change the risks resulting from floods. While some activities can be designed to mitigate the effects of flooding, many current practices and structures have unwittingly increased the flood risk.

In a humid climate during a major flood, a considerable part of the flow of a stream with a wide floodplain is carried by that floodplain. Clearing the floodplain for agriculture permits a progressively higher percentage of a large flood discharge to be carried by the floodplain. Some parts of the floodplain are eroded and other parts are built up by deposition of coarse sediment, while the channel capacity of the river channel is gradually reduced.

Drainage and irrigation ditches, as well as water diversions, can alter the discharge into floodplains and the channel's capacity to carry the discharge. The effects of agricultural and crop practices vary and depend upon the local soils, geology, climate, vegetation, and water management practices. In many countries agriculture dominates the use of land on floodplains. Where floods are seasonal, crops may be selected that can withstand floods of short duration and low volume during the flood season. Less resistant crops may be grown in the nonflood season.

Forest vegetation in general increases rainfall and evaporation while it absorbs moisture and lessens runoff. Deforestation or logging practices will reduce the vegetation and a forest's absorption capacity, thus increasing runoff. Overgrazing in grassland or rangeland areas decreases the vegetation cover and exposes soil to erosion as well as increased runoff. Cropland development may or may not increase runoff, depending on the land's prior use and the type of cropping patterns utilised.

Large dams affect the river channel both upstream and downstream from the dam and reservoir. Evaporation increases as a result of the expanded surface area of the reservoir, and this process tends to degrade the water quality. The reservoir acts as a sediment trap and the channel below the dam will regrade itself to accommodate the change in sediment load. The water, now with little sediment, scours the downstream channel. Dams may also increase ground-water recharge. They may raise the water table and even induce ground-water discharge into adjacent channels, thereby modifying stream discharge rates. Catastrophic dam failure produces a rapid loss of water from the reservoir and an instantaneously severe and dramatic change downstream.

Urbanisation of a floodplain or adjacent areas and its attendant construction increases runoff and the rate of runoff because it reduces the amount of surface land area available to absorb rainfall and channels its flow into sewers and drainage ways much more quickly. In short, floodplain dynamics are basic considerations to be incorporated in an integrated development planning study. It is essential that the study recognise that changes brought on by development can and will affect the floodplain in a multitude of ways. Early review of available flood hazard information and the programming of complementary flood hazard assessments are prudent and allow the planner to foresee and evaluate potential problems related to river hydraulics and floodplain dynamics. Then, mitigation measures can be identified to avoid or minimise these hazards and can be incorporated into the formulation of specific sectoral investment projects.

4.2. Satellite Remote Sensing Technology

Remote sensing technology can be especially useful and desirable when applied during the planning process. With remote sensing methods, the extent of floodplains and flood-prone areas can be approximated at small to intermediate map scales (up to 1:50,000) over entire river basins. Flood hazard maps can be prepared early in a development planning study to aid in defining and selecting mitigation measures for proposed sectoral development projects. In addition to discerning the risks of flooding, the same satellite data can be used to assess other hydrologic and atmospheric hazards as well as geologic and technological hazards. Furthermore, this satellite information can provide natural resource and land-use information at a small incremental cost once the basic data (computer compatible tapes [CCTs] or film image positives or negatives) are acquired.

It must be emphasized, however, that remote sensing technology is a tool, one of many that are employed by planners today. Application of this technology does not solve problems, but it can provide a planning study with recent, historical, and repetitive information.

4.2.1. Determining Acceptable Risk

Delineating floodplains and other areas subject to flooding is valuable input for proposing compatible development activities. Failure to understand the nature of flood hazards and to comprehend that they are not necessarily random in time and space, but are in fact roughly predictable conforming to statistical probability, can bring about increased flood risk. The planner should seek the contribution of a variety of disciplines to assess the risk of proposed activities.

Development planners need to know how often, on the average, the flood plain will be covered by water, for how long, and at what time of year. Natural changes as well as changes brought on by development activities affect the floodplain and must be understood to identify appropriate development and natural resource management practices. Changes in floodplain utilisation-such as urbanisation and more intensive agricultural production-can increase runoff and subsequent flood levels. It is critical for the planner to appreciate these and other effects of land-use change. Early consultation with water resource and management specialists during the planning study is prudent, for it enables the planner to foresee and evaluate potential conflicts between present and proposed land use and their relationship to flood events and the hazards they may pose.

Acceptable risk criteria can help in distinguishing between different degrees of risk for different development activities and in evaluating constraints associated with potential investment projects. The chosen acceptable frequency of a particular flood event should be appropriate for the type of development activity. For example, it may well be worth the risk of occasional flooding to plant crops in the floodplain where soils are enriched by cyclical flooding and the deposition of sediments. Resulting sand and gravel deposits may lead to commercial exploitation. On the other hand, it is more appropriate to site a large agroindustrial or housing project in an area with a very small probability of a large flood occurring each year.

4.2.2. Satellite Remote Sensing to Flood Hazards

Floodplain mapping techniques are either dynamic or static methods. Many traditional techniques are dynamic: they monitor the continuous change in river or stream flow and require considerable field work and maintenance of long-term records. Some traditional dynamic techniques utilise regression analysis and rainfall estimates derived from models in which long-term records are transferred from similar basins or reaches in a given region. Though these methods do require the application of some records, they may be used where long-term records do not exist for the particular stream or river under study. In any event, the principal objectives of using dynamic techniques are to calculate the return period or frequency of particular flood events and to determine stream flow and flood-level characteristics. These are important for the planner to know in order to adequately weigh the risk of development in a floodplain.

Flood inundation and floodplain maps have been prepared from satellite data for more than a decade by hydrologists all over the world. These are considered static techniques since they characterise the area at a particular point in time. While a dynamic long-term flood history is desirable, such static techniques are capable of yielding useful information for flood hazard assessment, especially in the diagnostic and preliminary stages of an integrated development planning study.

In the absence of information from dynamic techniques, it is possible to estimate the probability of a flood event occurrence when information from static techniques is combined with historical flood observations, disaster reports, and basic natural resource information, particularly hydrologic data. Flood event frequency estimates, particularly for an extreme event, is valuable information to the planning study.

While inexpensive photo-optical processing techniques of satellite data are still valid, the increasing price and decreasing availability of film imagery, and innovative use of digital-to-analog data processing, make computer-assisted analysis a viable option. The commonly used Landsat Multispectral Scanner (MSS) data and the high-resolution Landsat Thematic Mapper (TM) and SPOT High Resolution Visible Range (HRV) data with the potential for larger scale mapping are examples. Also, the small-scale resolution but synoptic regional coverage provided by the NOAA satellite series carrying the Advanced Very High Resolution Radiometer (AVHRR) provides a highly informative aid to planners in determining the extent of flood events.

One of the requirements of an integrated development planning study is to develop a clear definition of the study area and a sense of the region's general development situation. The relationship of the region's natural goods, services, and its hazards and current natural resource management practices should be put in the context of affected ecosystems.

In order to integrate flood plain information into a planning study, the definition of floodplains and flood-prone areas and the probability of a given event occurring during the lifetime of a development project should be determined. This information will assist in making decisions about whether or not a certain level of risk is acceptable. It is important to bear in mind that floodplain and flood hazard maps are not intended to be substitutes for, but rather precursors to, engineering design studies.

A variety of mitigation measures can be identified and selected which will reduce or minimise the impact of flooding. Such mitigation measures include adopting land-use classification and zoning systems, building codes, taxation, and insurance programs, in addition to the prevalent "user beware" approaches. Remote sensing technology can and should play an important role in the design of the planning study. Map scales of collected information will no doubt vary. Small scale satellite image maps complement traditional thematic maps with synoptic spatial information that can be used as a basis for a regional assessment of the hydrologic regimen, including floodplain definition for major river valleys. Indeed, state-of-the-art technology now permits preparation of thematic image map within U.S. national map accuracy standards for scales as large as 1:50,000.

4.3. Flood Mitigation Measures

— *Modify Hazard:* dams, catchment areas, retention ponds, high flow diversions, cropping patterns, reforestation.

— *Modify Water Courses:* dikes, levees, channeling, stream rectification, erosion control, drainage systems.

— *Modify Structures:* building elevation or strengthening, flood proofing.

— *Modify Land Use:* use zones, subdivision regulations, sanitary and water-well regulations, development restrictions, easements and setbacks, floodplain management taxation.

— *Insurance:* flood insurance programs.

— *Forecast, Warning and Emergency Systems:* flood monitoring, alert systems, evacuation and rescue plans, shelter and emergency relief.

4.4. Flood Hazard Mapping Techniques

Traditionally, gathering and analysing hydrologic data related to floodplains and flood-prone areas has been a time-consuming effort requiring extensive field observations and calculations. This traditional approach uses historical data of flood events to delineate the extent and recurrence interval of flooding.

With the development of remote sensing and computer analysis techniques, now traditional sources can be supplemented with these new methods of acquiring quantitative and qualitative flood hazard information. This static approach uses indicators of flood susceptibility to assess an area's flood proneness. Both of these approaches are discussed below.

4.4.1. Traditional Techniques of Floodplain Mapping

Conventional dynamic flood frequency analysis techniques have been developed to quantitatively assess flood hazards over the past half century. These traditional techniques yield dynamic historical flood data which, when available, is used to accurately map floodplains. In addition to a record of peak flows over a period of years (frequency analysis), a detailed survey (cross sections, slopes and contour maps) along with hydraulic roughness estimates is required before the extent of flooding for an expected recurrence interval can be determined. In traditional floodplain mapping, the requisite data and maps include the following:

— The selected base (topographic) map with the surface water system

— Hydrologic data:
  — Frequency analysis (including river discharge and historical flood data)
  — Flood inundation maps
  — Flood frequency and damage reports, etc.
  — Stage-area curves
  — Slope maps
  — Cross sections
  — Hydraulic roughness
— Related maps such as soils, physiography, geology, hydrology, land use, vegetation, population density, infrastructure, and settlements.

This dynamic approach requires extensive long term field surveys, with a network of gauging stations that can develop the data needed for precise risk assessments. Such extensive long term information is seldom available for river systems in less developed countries. To obtain hydrologic data, one must contact the appropriate hydrometeorological agencies of government to secure available data and maps. Soils maps and geological maps often delineate floodplains. Topographic maps at suitable scales for the project should be available within the country. What is more readily available is information derived from static techniques which are capable of yielding information on flood hazard assessment.

4.4.2. Remote Sensing Techniques for Floodplain Mapping

For large areas, such as major river valleys, time and funds available are often limited. Therefore, it is usually not possible to conduct expensive detailed hydrologic data gathering, analysis, and mapping activities during a planning study. Remote sensing technology, especially space technology, now provides an economically feasible alternative means of supplementing traditional hydrologic data sources. These static techniques provide pictures of an area that can be analysed for certain flood-related characteristics and can be compared to images from an earlier or later date to determine changes in the study area.

Remote sensing methods require a platform such as a satellite (e.g., Landsat) or an aircraft, plus a sensor such as an MSS on the platform. Satellite imagery can be acquired in digital (CCT) or analog (film) formats. Digital data may not be an alternative because of the expense and requirement for sophisticated computer hardware and software. Therefore, the focus of the method presented here is to provide a technique which uses original or raw film data for floodplain mapping and floodplain hazard assessments. The concept of preprocessing CCTs is also discussed below since it is feasible to acquire digitally enhanced film products for these applications.

Flood-inundation and flood hazard maps have been prepared by many hydrologists all over the world from aircraft and satellite data, mostly from the visible and infrared bands. A few hydrologists have used thermal infrared data to map flooded areas.

Satellite data can be used to find indicators of floodplains, and may be easier to use than aircraft images in delineating floodplains. Computer-enhanced information from aerial photography or a combination of this with satellite imagery has been used. Digitised color-infrared aerial photographs to classify vegetation that correlates with floodplains have also been used. Landsat digital data have also been combined with digital elevation data to develop stage-area relationships of flood-prone areas. A comprehensive reference for satellite remote sensing techniques relating to water resources is Satellite Hydrology, which has more than 100 papers on the subject.

4.4.3. Landsat Floodplain Indicators

- Upland physiography;
- Watershed characteristics such as shape, drainage density, etc.;
- Degree of abandonment of natural levees;
- Occurrence of stabilised sand dunes on river terraces;
- Channel configuration and fluvial geomorphic characteristics;
- Backswamp areas;
- Soil-moisture availability (also a short-term indicator of flood susceptibility);
- Soil differences;
- Vegetation differences;
- Land-use boundaries;
- Agricultural development; and
- Flood alleviation measures on the floodplain.

Floods, hydraulic forces, engineering structures, and development on the floodplain can and do result in physical changes in the river channel, sedimentation patterns, and flood boundaries. It is very costly to continually update maps to accurately depict these changing conditions. Satellite imagery can provide a record of changes to complement maps and conventional point source data. Hence, up-to-date satellite imagery of the study area can be compared with previously collected data to determine changes during specific time periods. Similarly, in mapping a flood using satellite imagery, the inundated area can be compared with a map of the area under preflood conditions.

The flood often leaves its imprint or "signature" on the surface in the form of soil moisture anomalies, pounded areas, soil scours, stressed vegetation, debris lines, and other indicators of the flooded area for days, or even weeks, after the flood waters have receded. It should be noted that delineation of floodplains using remote sensing data cannot, by itself, be directly related to any return period. However, when it is used in conjunction with other information, the delineated floodplain can be related to an estimated or calculated event. This static method can reveal an area's flood proneness and yield information useful for a flood hazard assessment.

4.4.4. Selection of Satellite Data

A critical but generally underestimated requirement for effective use of satellite imagery in flood hazard assessments is the selection of data. A number of sensors on board Earth observation satellites have provided data suitable for mapping floodplains and areas inundated by floods. The sensing systems and observation satellites which have been in operation for the longest period of time are the MSS on all five of the Landsat series and the AVHRR on the current NOAA satellite series. More recent sensing systems and satellites include the TM on Landsat 4 and 5 and the SPOT satellite with HRV sensors. Each system has its spatial, spectral, and temporal advantages and limitations.

Other remote sensing systems such as those found on the U.S. Nimbus and Seasat satellites and the Space Shuttle have been used experimentally, but their coverage is sporadic. Landsat, NOAA, and SPOT satellites collect data in a digital mode. The data products can be purchased as CCTs or in analog form as photographic prints or film transparencies. SPOT and Landsat program film product costs are such that the cost of producing thematically enhanced photo-optical data products for specific applications such as floodplain delineation and flood mapping now approaches the cost of digital image processing.

4.5. Advantages and Limitations of Satellite for Flood Hazard Assessments

— *Landsat MSS:* provides data for relatively small-scale mapping (1:1,000,000 -1:100,000), with coverage only once every 16 days in 4 spectral bands.

— *Landsat TM:* data are collected with the same frequency as MSS data in six of seven solar reflective spectral bands (1,2,3,4,5, and 7) and is suitable for larger scale mapping (up to 1:50,000).

— NOAA AVHRR; provides multispectral coverage four times each day (two daytime and two nighttime) but produces data adequate for only small-scale mapping (1:3,000,000 -1:500,000); most useful in delineating maximum flood coverage of surface areas.

— *Spot HRV:* the SPOT satellite HRV sensors provide data for relatively large-scale mapping (up to 1:25,000) from three spectral bands once every 26 days. It has a pointable sensor mode that can provide data on a more frequent basis.

One limitation found in all of the above sensors is that none provide cloud penetration, which may limit the amount of data available in humid, cloud-covered areas. Since most satellite coverage for a single full scene extends over a large area (usually more than 33,000km^2, except for a SPOT scene, which covers approximately 3,600km^2), advantages and requirements of each system are important to keep in mind.

In deciding on the scale of the base map for the study, which is dependent on the scale of available topographic maps, it is of primary importance to consider the potential use of satellite data.

4.5.1. Photo-optical Method for Initial Floodplain Delineation and Flood Hazard Assessment

Integrated regional development planning studies do not traditionally include original flood hazard assessments but rather depend on existing, available information. If time and budget constraints do not permit a detailed, large-scale assessment to be carried out, a floodplain map and a flood hazard assessment can be prepared using the photo-optical method, using Landsat data and the planning study information which is usually available.

In mapping floodplains, black-and-white positive film transparencies of Landsat imagery in 70mm format are especially useful for floodplain delineation. Applicable map scales range from 1:1,000,000 to 1:100,000 or larger, depending on the availability of complementary flood hazard assessment and hydrologic information. Their usefulness is achieved through their analysis with a color-additive viewer, which provides the greatest capability and flexibility for optical multispectral (more than one band), multitemporal (scenes from two dates), and multiscale (images with different scales) analysis. If 70mm film products are not available, film positives at 1:1,000,000 can be cut or reduced to 70mm size and used in the color-additive viewer. This technique permits the imagery to be used as a base for producing enlargements of subscenes.

4.5.2. Advantages of using Landsat Data

— Flexibility in using either film-positive transparencies or CCTs purchased directly from the satellite data distributor.

— Flexibility in using either a color-additive viewer, photographic laboratory, or computer for image processing, analysis, and composting.

— Ability to concurrently use scenes from two dates for comparing pre-event, event, and post-event situations.

— Flexibility in producing 35mm slides, photographic prints, or film-positive transparencies for use at selected base-map scales.

The photo-optical data-processing method described above has been developed as a low-cost alternative to digital image processing. Digital image processing requires expensive multispectral image analysers, computers, film writers and appurtenant equipment in addition to a custom photographic laboratory.

While data prices vary from source to source and country to country, experience has shown that the per square kilometer cost of data acquisition, analysis, and preparation of analog products may range from U.S. 4 cents using film-positive transparency data format to U.S. 20 cents for CCT data format. A remote sensing specialist familiar with photo-optical or computer-enhanced multispectral analysis systems, in collaboration with other planning studies and with regional complementary information and logistical support, would be able to carry out a flood hazard assessment and prepare a flood plain map for a 30,000-90,000 km^2/area at a scale of up to 1:250,000 in approximately a one month time period. Exact time allocation obviously depends on the scale of the final map product to be produced, the density of the surface water system, the topography, and the availability of relevant natural resource and infrastructure maps at appropriate scales.

Many countries do have a color-additive viewer available for photo-optical analysis. Most planning, water, and natural resource agencies, however, do not have adequate funds or the need for a full-time dedicated digital image processing facility for computer-assisted map analysis. If use of such technology is desired, international state-of-the-art digital data processing facilities are recommended. Access to either analysis system can be facilitated by specialists who are familiar with satellite data sources; the selection of available imagery, its purchase and processing; and the analysis of analog products.

A facility equipped with only photo-optical equipment and access to a photographic laboratory can utilise digital image processing by arranging for preprocessing CCTs at a qualified facility. Raw data and enhanced film products can be produced on a custom basis for specific applications and be made in formats compatible with the photo-optical equipment available to the user. Such processing should be performed, if at all possible, by a photograph developing and printing specialist in collaboration with a computer programmer and professionals knowledgeable about the study area.

Conversion from a digital to analog or film mode at an early stage of a project will eliminate the need for a dedicated computer capability at many institutions and at the same time can increase the efficiency of the selected digital image processing facility. The film products produced from the digital analysis can then be effectively and efficiently used in the photo-optical data systems of the user without the need for such photographic reprocessing as contrast enhancements, film-density balancing, and extensive black-and-white and color film development and printing. The value and effectiveness of equipment such as a color-additive viewer is actually increased, since digitally enhanced and corrected imagery will be used instead of raw data.

The repetitive coverage of any area by operational Earth observation satellites makes it possible to monitor dynamic features of flooding that can cause changes, e.g, changes in the channel of the river itself or floodplain boundaries. Further, the spatial distribution of the features that have changed can be readily mapped by techniques of temporal analysis developed since the launch of Landsat 1 in 1972. Slides of full scenes and subscenes can be projected at any scale for analysis. The slides can be projected onto a base map, thematic maps, and enlarged satellite single-band prints to produce thematic image maps.

4.5.3. Advantages of Digital Image Processing

— Automatic spatial measurements.

— Thematic enhancements, such as linear contrast stretches, band rating, geometric and atmospheric corrections, edge enhancements, etc.

— Maximum scene processing versatility.

— Potential utilisation in geographic information systems.

5

Hurricane Hazards

All of the embryonic tropical depressions that develop into hurricanes originate in similar meteorological conditions and exhibit the same life cycle. The distinct stages of hurricane development are defined by the "sustained velocity" of the system's winds-the wind velocity readings maintained for at least one minute near the center of the System. In the formative stages of a hurricane, the closed isobaric circulation is called a tropical depression. If the sustained velocity of the winds exceeds 63km/h (39 mph), it becomes a tropical storm. At this stage it is given a name and is considered a threat. When the winds exceed 119km/h (74 mph), the system becomes a hurricane, the most severe form of tropical storm. Decay occurs when the storm moves into nontropical waters or strikes a landmass. If it travels into a nontropical environment it is called a subtropical storm and subtropical depression; if landfall occurs. the winds decelerate and it becomes again a tropical storm and depression.

5.1. Tropical Depression

Hurricanes are generated at latitudes of 8 to 15 degrees north and south of the Equator as a result of the normal release of heat and moisture on the surface of tropical oceans. They help maintain the atmospheric heat and moisture balance between tropical and non-tropical areas. If they did not exist, the equatorial oceans would accumulate heat continuously.

Hurricane formation requires a sea surface temperature of at least 27 degrees Celsius (81 degrees Fahrenheit). In the summer months, the sea temperatures in the Caribbean and Atlantic can reach 29 degrees (84 degrees), making them prime locations for inception. The surface water warms the air, which rises and then is blocked by warmer air coming from the easterly winds. The meeting of these two air masses creates an atmospheric inversion. At this stage, thunderstorms develop and the inversion may be broken, effectively lowering the atmospheric pressure.

5.2. Tropical Storm and Hurricane

The growth of the system occurs when pressure in the center of the storm drops well below 1000 millibars (mb) while the outer boundary pressure remains normal. When pressure drops, the trade winds are propelled in a spiral pattern by the earth's rotation. The strong torque forces created by the discrepancy in pressure generate wind velocities proportional to gradient of pressure. As the energy level increases, the air circulation pattern is inward towards the low pressure center and upward, in a counter-clockwise spiral in the northern hemisphere and clockwise in the southern hemisphere. The cycle perpetuates itself and the organised storm begins a translational movement with velocities of around 32 km/h during formation and up to 90km/h during the extra-tropical life.

The zone of highest precipitation, most violent winds, and rising sea level is adjacent to the outer wall of the "eye." The direction of the winds, however, is not towards the eye but is tangent to the eye wall about 50km from the geometric center. The organised walls of clouds are composed of adjoining bands which can typically reach a total diameter of 450km. The central eye, unlike the rest of the storm, is characterised as an area of relatively low wind speeds and no cloud cover with an average diameter of 50-80km and a vertical circulation of up to 15km.

Hurricane classification is based on the intensity of the storm, which reflects damage potential. The most commonly used categorisation method is the one developed by H. Saffir and R.H. Simpson. The determination of a category level depends mostly on barometric pressure and sustained wind velocities. Levels of storm surge fluctuate greatly due to atmospheric and bathymetric conditions. Thus, the expected storm surge levels are general estimates of a typical hurricane occurrence.

5.3. Landfall or Dissipation

Typically, a hurricane eventually dissipates over colder waters or land about ten days after the genesis of the system. If it travels into a non-tropical environment, it loses its energy source and falls into the dominant weather pattern it encounters. If, on the other hand, it hits land, the loss of energy in combination with the increased roughness of the terrain will cause it to dissipate rapidly. When it reaches land in populated areas, it becomes one of the most devastating of all natural phenomena.

5.4. Temporal Distribution of Hurricane Occurrence in the Caribbean

The official hurricane season in the Greater Caribbean region begins the first of June and lasts through

November 30, with 84 percent of all hurricanes occurring during August and September. The greatest risk in Mexico and the western Caribbean is at the beginning and end of the season, and in the eastern Caribbean during mid-season.

Every year over 100 tropical depressions or potential hurricanes are monitored, but an average of only ten reach tropical storm strength and six become hurricanes. These overall averages suggest that activity is uniform from year to year but historical records indicate a high degree of variance, with long periods of tranquillity and activity. The Atlantic basin has the widest seasonal variability. In 1907, for example, not a single tropical storm reached hurricane intensity, while in 1969, there were 12 hurricanes in the northern Atlantic.

Because the cycles vary in periodicity and duration, prediction is difficult. Recent forecasting developments, connecting hurricane activity levels with El Niño and the Quasi-biennial Oscillation have made it possible to predict the variance in Atlantic seasonal hurricane activity with an accuracy of 40 to 50 percent, but this degree of accuracy, while considered high by meteorological standards, is not good enough for planners trying to develop appropriate emergency response systems.

5.4.1. Hazardous Characteristics of Hurricanes

5.4.1.1. Winds

Hurricane wind speeds can reach up to 250km/h (155mph) in the wall of the hurricane, and gusts can exceed 360km/h (224mph).The destructive power of wind increases with the square of its speed. Thus, a tripling of wind speed increases destructive power by a factor of nine. Topography plays an important role: wind speed is decreased at low elevations by physical obstacles and in sheltered areas, while it is increased over exposed hill crests. Another contributor to destruction is the upward vertical force that accompanies hurricanes; the higher the vertical extension of a hurricane, the greater the vertical pulling effect.

Destruction is caused either by the direct impact of the wind or by flying debris. The wind itself primarily damages agricultural crops. Entire forests have been flattened by forces that pulled the tree roots from the earth. Man-made fixed structures are also vulnerable. Tall buildings can shake or even collapse. The drastic barometric pressure differences in a hurricane can make well-enclosed structures explode and the suction can lift up roofs and entire buildings. But most of the destruction, death, and injury by wind is due to flying debris, the impact force of which is directly related to its mass and the square of its velocity. The damage caused by a flying car to whatever it strikes will be greater than if the wind had acted alone. Improperly fastened roof sheets or tiles are the most common projectiles. Other frequent objects are antennas, telephone poles, trees, and detached building parts.

Building standards to withstand high wind velocities are prescribed in almost all countries that face a high risk. The codes recommend that structures maintain a certain level of strength in order to withstand the local average wind velocity pressure, calculated by averaging wind pressure over a period of ten minutes for the highest expected wind speed in 50 years. The Caribbean Uniform Building Code (CUBIC) under consideration by the Caribbean countries, prescribes the reference wind velocity pressure for each country.

*5.4.1.2. **Rainfall***

The rains that accompany hurricanes are extremely variable and hard to predict. They can be heavy and last several days or can dissipate in hours. The local topography, humidity, and the forward speed of a hurricane in the incidence of precipitation are recognised as important, but attempts to determine the direct connection have so far proved futile.

Intense rainfall causes two types of destruction. The first is from seepage of water into buildings causing structural damage; if the rain is steady and persistent, structures may simply collapse from the weight of the absorbed water. The second, more widespread and common and much more damaging, is from inland flooding, which puts at risk all valleys along with their structures and critical transportation facilities, such as roads and bridges.

Landslides, as secondary hazards, are often triggered by heavy precipitation. Areas with medium to steep slopes become oversaturated and failure occurs along the weakest zones. Thus, low-lying valley areas are not the only sites vulnerable to precipitation.

5.4.1.3. Storm Surge

A storm surge is a temporary rise in sea level caused by the water being driven over land primarily by the on-shore hurricane force winds and only secondarily by the reduction in sea-level barometric pressure between the eye of the storm and the outer region. Another estimate is that for every drop of 100 millibars (mb) in barometric pressure, a 1m (3.28 feet) rise in water level is expected. The magnitude of the surge at a specific site is also a function of the radius of the maximum hurricane winds, the speed of the system's approach, and the foreshore bathymetry. It is here that the difficulty arises in predicting storm surge levels. Historical records indicate that the increase in mean sea level can be negligible or can be as much as 7.5 meters (24.6 feet). The most vulnerable coastal zones are those with the highest historical frequencies of landfalls. Regardless of its height, the great dome of water is often 150km (93 miles) wide and moves toward the coastline where the hurricane eye makes landfall.

Storm surges present the greatest threat to coastal communities. Ninety percent of hurricane fatalities are due to drowning caused by a storm surge. Severe flooding from a storm surge affects low-lying areas up to several kilometers inland. The height of the surge can be greater if man-made structures in bays and estuaries constrict water flow and compound the flooding. If heavy rain accompanies storm surge and the hurricane landfall occurs at a peak high tide, the consequences can be catastrophic. The excess water from the heavy rains inland creates fluvial flooding, and the simultaneous increase in sea level blocks the seaward flow of rivers, leaving nowhere for the water to go.

5.4.1.4. Hurricane Gilbert

Hurricanes are by far the most frequent hazardous phenomena in the Caribbean. Tomblin states that in the last 250 years the West Indies has been devastated by 3 volcanic eruptions, 8 earthquakes, and 21 major hurricanes. If tropical storms are also taken into account, the Greater Caribbean area has suffered from hundreds of such events.

The economic and social consequences of this phenomenon are severe, especially in less developed countries, where a significant percentage of the GDP can be destroyed by a single event. Without a comprehensive list of costs and casualties, the economic and social disruption caused by a disastrous event is hard to grasp.

Hurricane Gilbert struck the Caribbean and the Gulf Coast of Mexico in 1988, causing comprehensive damage in Mexico, Jamaica, Haiti, Guatemala, Honduras, Dominican Republic, Venezuela, Costa Rica, and

Nicaragua. Arriving in Saint Lucia as a tropical depression, it resulted in damage estimated at US$2.5 million from the flooding and landslides caused by the heavy rain.

The physical variations in this hurricane resulted in different types of damage. It was considered a "dry" hurricane when it struck Jamaica, discharging less precipitation than would be expected. Thus, most of the damage was due to wind force which blew away roofs. By the time it approached Mexico, however, it was accompanied by torrential rains, which caused massive flooding far inland.

Hurricane Gilbert began as a tropical wave on September 3, 1988, on the north coast of Africa. Six days later, the system was across the Atlantic and had evolved into Gilbert as a tropical storm. It struck Jamaica on September 12 as a Category 3 (SSH Scale) hurricane and traveled westward over the entire length of the island. Gaining strength as it moved northwest, it hit the Yucatan Peninsula, in Mexico, on September 14, as a Category 5 (SSH Scale) hurricane. By September 16 it had been weakened and finally dissipated after moving inland over the east coast of Mexico.

Sustained winds in Jamaica were measured at 223 km/h, and greater across high ridges. The barometric pressure was the lowest ever recorded in the Western Hemisphere at 888mb, 200km east-southeast of Jamaica. The barometric pressure when it hit Jamaica was 960mb. The forward speed was 31 km/h. The eye had a 56km diameter, but little storm surge occurred in Jamaica. Average rainfall registered from 250mm to 550mm. Serious flooding due to storm surge and heavy rains was not a problem. Landslides occurred at high elevations where most of the rainfall was concentrated.

By the time Hurricane Gilbert hit Mexico it had changed characteristics. In the Yucatan the storm surge reached 5 meters in height and rainfall averaged 400mm. By the time Gilbert struck the northern coast of Mexico, the winds had increased to 290km/h and the storm surge had reached 6 meters.

Affected population and damage to social sectors

Even though the loss of life was limited to 45 reported deaths, 500,000 people lost their homes when approximately 280,000 houses-almost 55 percent of the housing stock-were damaged in Jamaica. Of these, 14,000, or 5 percent, were totally destroyed and 64,000 were seriously damaged.

The Government of Mexico reported that the hurricane caused 200 deaths and approximately 200,000 homeless. In the state of Nuevo Leon, where the Monterrey area suffered from extensive flooding, 100 people died and 30,000 housing units were destroyed.

Impact on the economy and damage to productive sectors

The Planning Institute of Jamaica estimated the total direct damage at US$956 million. Nearly half was accounted for by losses from agriculture, tourism, and industry; 30 percent from housing, health, and education infrastructure; and 20 percent from economic infrastructure. The economic projections for 1988 had to be adjusted dramatically, to allow for expected losses of US$130 million in export earnings, and more than US$100 million in tourism earnings; therefore, instead of the expected 5 percent growth in GDP, a decline of 2 percent was projected. Other estimates were for increases in inflation (30 percent), government public expenditures (US$200 million), and public sector deficit (from 2.8 percent to 10.6 percent of GDP).

As expected, the economic activity most affected was agriculture, with the total destruction of banana and broiler production and of more than 50 percent of the coffee and coconut crops. Capital losses to the sector were estimated at J$0.7 billion. According to some calculations, the loss of revenue through 1992 will be US$214 million.

Other productive sectors were also affected seriously. Manufacturing suffered J$600 million (in 1989 dollars) in losses, mainly from a decline of 12 percent in its exports. Tourism lost US$90 million in foreign exchange, with 5 percent fewer visitor arrivals in the third quarter of 1988 than during the same time period in 1987. Loss of electricity decreased bauxite production by 14.2 percent for that quarter compared to the third quarter of the previous year, and alumina exports declined by 21 percent.

The tourism industry suffered the greatest damage. The tourist areas of the state of Quintana Roo, for example, suffered US$100 million in direct damage and lost an estimated US$90 million in revenues. The Inter-American Development Bank, after evaluating the damage to infrastructure in this sector, lent US$41.5 million for reconstruction.

Damage to natural resources

The coastal resources of Jamaica suffered extensive damage from hurricane forces. It is estimated that 50 percent of the beaches were seriously eroded, with the northeast coast being the most affected. An estimated 60 percent of all the trees in the mangrove areas were lost, 50 percent of the oyster culture was unsalvageable, and other non-measurable harm occurred to coral reefs and the water quality of the island.

The impact across the Yucatan Peninsula in terms of damage to wildlife, beaches, and coral reefs was much higher than on the coasts of Jamaica. Extensive reduction in beaches and coral reefs was reported, and large numbers of birds lost their lives.

5.4.2. Risk Assessment and Disaster Mitigation

The risk posed by hurricanes to a particular country is a function of the likelihood that a hurricane of a certain intensity will strike it and of the vulnerability of the country to the impact of such a hurricane. Vulnerability is a complex concept, which has physical, social, economic and political dimensions. It includes such things as the ability of structures to withstand the forces of a hazardous event, the extent to which a community possesses the means to organise itself to prepare for and deal with emergencies, the extent to which a country's economy depends on a single product or service that is easily affected by the disaster, and the degree of centralisation of public decision-making.

Population centers and economic activities in the region are highly vulnerable to disruption and damage from the effects of extreme weather. They are largely concentrated in coastal plains and low-lying areas subject to storm surges and landborne flooding. High demands placed on existing lifeline infrastructure, combined with inadequate funds for the expansion and maintenance of these vital systems, have increased their susceptibility to breakdowns. Uncontrolled growth in urban centers degrades the physical environment and its natural protective capabilities. Building sites safe from natural hazards, pollution, and accidents have become inaccessible to the urban poor, who are left to build their shelters on steep hillsides or in flood-prone areas. Agriculture, particularly the cultivation of bananas for export, is often practiced without the necessary conservation measures corresponding to the soil, slope, and rainfall characteristics of the area.

Communities, countries, or regions differ greatly in vulnerability, and hence in the effects they may suffer from hurricanes of similar strength. The very size of a country is a critical determinant of vulnerability: small island nations can be affected over their entire area, and major infrastructure and economic activities may be crippled by a single event. Scarce resources that were earmarked for development projects have to be diverted to relief and reconstruction, setting back economic growth.

To assess future risks, planners must study historical trends and correlate them with probable future changes. The main cause of increasing vulnerability is the population movement to high-risk areas. Most cities in the West Indies are in low coastal zones threatened by storm surge, and they continue to grow.

The economic sectors most affected by hurricanes are agriculture and tourism. Together, these represent a major portion of the economy for the countries in the Caribbean. Particularly for island countries, agriculture is the most vulnerable activity. Hurricanes have disastrous effects on banana crops in particular. During Hurricane Alien, in August of 1980, Saint Lucia suffered US$36.5 million in damage, with 97 percent of the banana plantations destroyed. In St. Vincent 95 percent, and in Dominica 75 percent, of the banana plantations were ruined. Damage to the tourism industry is more difficult to quantify since it includes many other economically identifiable sectors such as transportation and hotel services.

Crop statistics rarely account for long-term losses. The increased salinity in the soil due to the storm surge can have detrimental effects on production in subsequent years. For example, Hurricane Fifi decreased production in Honduras by 20 percent the year it occurred, but in the following year production was down by 50 percent. How much of this reduction was due to the increase in salinity is unclear, but it is known that salt destroys vegetation slowly.

5.4.3. Mitigation Measures

Once the risk posed by hurricanes is understood, specific mitigation measures can be taken to reduce the risk to communities, infrastructure, and economic activities. Human and economic losses can be greatly reduced through well-organised efforts to implement appropriate preventive measures, in public awareness and in issuing timely warnings. Thanks to these measures, countries in the region have experienced a drastic reduction in the number of deaths caused by hurricanes.

Mitigation measures are most cost-effective when implemented as part of the original plan or construction of vulnerable structures. Typical examples are the application of building standards designed for hurricane-force winds, the avoidance of areas that can be affected by storm surge or flooding, and the planting of windbreaks to protect wind-sensitive crops. Retrofitting buildings or other projects to make them hurricane-resistant is more costly and sometimes impossible. Once a project is located in a flood-prone area, it may not be feasible to move it to safer ground.

The overall record on mitigation of hurricane risk in the Caribbean and Central America is not very encouraging. Cases abound of new investments in the public or productive sectors that were exposed to significant hazard risk because of inappropriate design or location, and even of projects that were rebuilt in the same way on the same site after having been destroyed a first time. Other cases can be cited of schools and hospitals funded with bilateral aid that were built to design standards suitable for the donor country but incapable of resisting hurricane-strength winds prevalent in the recipient country.

The tourism sector in the Caribbean is notorious for its apparent disregard of the risk of hurricanes and associated hazards. A hotel complex built with insufficient setback from the high-water mark not only risks being damaged by wave action and storm surge, but interferes with the normal processes of beach formation and dune stabilisation, thus reducing the effectiveness of a natural system of protection against wave action. After the first serious damage is incurred the owners of the hotel will most likely decide to rebuild on the same site and invest in a seawall, rather than consider moving the structure to a recommended setback.

In the past three decades the technological capacity to monitor hurricanes has improved dramatically, and along with it the casualty rate has declined. New technology permits the identification of a tropical depression and on-time monitoring as the hurricane develops. The greatest advance has occurred in the United States, but developing countries benefit greatly because of the effective warning mechanism. The computer models also generate vast quantities of information useful for planners in developing nations.

Computer models that estimate tracking, landfall, and potential damage were first implemented in 1968 by the U.S. National Hurricane Center (NHC). At this point there are five operational track guidance models: Beta and Advection Model (BAM), Climatology and Persistance (CLIPER), a Statistical-dynamical model (NHC90), Quasi-Lagrangian model (QLM) and the barotropic VICBAR. They vary in capacity and methodology and occasionally result in conflicting predictions, though fewer than formerly. The NHC evaluates incoming data on all tropical storms and hurricanes in the Atlantic and eastern Pacific tropical cyclone basin and issues an official track and intensity forecast consisting of center positions and maximum one-minute wind speeds for 0, 12, 24, 48, and 72 hours.

The NHC has also developed a hurricane surge model named Sea, Lake and Overland Surges (SLOSH) to simulate the effects of hurricanes as they approach land. Its predecessor SPLASH, used in the 1960s, was useful for modeling hurricane effects along smooth coastlines, but SLOSH adds to this a capability to gauge flooding in inland areas. These results can be used in planning evacuation routes.

A computerised model that assesses the long-term vulnerability of coastal areas to tropical cyclones has also been developed. This model, the National Hurricane Center Risk Analysis Program (HURISK), uses historical information on 852 hurricanes since 1886. The file contains storm positions, maximum sustained winds, and central pressures (unavailable for early years) at six-hour intervals. When the user provides a location and the radius of interest, the model determines hurricane occurrences, dates, storm headings, maximum winds, and forward speeds. Vulnerability studies begin when the median occurrence date, direction distribution, distribution of maximum winds, probability of at least x number of hurricanes passing over n consecutive years, and gamma distribution of speeds are determined. Planners can use these objective return period calculations to evaluate an otherwise subjective situation.

One of the most important steps a country can take to mitigate the impact of hurricanes is to incorporate risk assessment and mitigation measure design into development planning. The design of basic mitigation measures begins with the compilation of all historical records of former hurricane activity in the country, to determine the frequency and severity of past occurrences. Reliable meteorological data for each event, ranging from technical studies to newspaper reports, should be gathered. With all the information in place, a study of (1) the distribution of occurrences for months of a year, (2) frequencies of wind strengths and direction, (3) frequencies of storm surges of various heights along different coastal sections, and (4) frequencies of river flooding and their spatial distribution should be undertaken. The statistical analysis should provide quantitative support for planning strategies.

The design of mitigation measures should follow the statistical analysis and consider long-term effects. Both non-structural and structural mitigation measures should be considered, taking into account the difficulties of implementation.

Non-structural measures consist of policies and development practices that are designed to avoid risk, such as land use guidelines, forecasting and warning, and public awareness and education. Much credit for the reduction of casualties from hurricanes in the Caribbean should be given to the Pan Caribbean Disaster Preparedness and Prevention Project (PCDPPP), which has worked effectively with national governments on motivating the population to take preventive measures, such as strengthening roof tie-downs, and on establishing forecasting and warning measures.

Structural mitigation measures include the development of building codes to control building design, methods, and materials. The construction of breakwaters, diversion channels, and storm surge gates and the establishment of tree lines are a few examples of mitigation from a public works standpoint.

The effectiveness of national emergency preparedness offices of countries in the region is often seriously limited because of inadequate institutional support and a shortage of technical and financial resources. In the smaller Caribbean islands, these offices are mostly one-person operations, with the person in charge responsible for many other non-emergency matters. It would be unrealistic to expect them to be able to act effectively at the local level in cases of area-wide emergencies, such as those caused by hurricanes. It is therefore essential to enhance the capacity of the population in small towns and villages to prepare for and respond to emergencies by their own means.

From 1986 through 1989, the OAS/Natural Hazards Project has worked with several Eastern Caribbean countries to evaluate the vulnerability of small towns and villages to natural hazards, and train local disaster managers and community leaders in organising risk assessments and mitigation in their communities. These activities have resulted in the preparation of a training manual with accompanying video for use by local leaders. This effort has focused on lifeline networks-transportation, communications, water, electricity, sanitation-and critical facilities related to the welfare of the inhabitants, such as hospitals and health centers, schools, police and fire stations, community facilities, and emergency shelters.

5.4.4. Surviving Damage and Disruption

The degree to which local communities can survive damage and disruption from severe storms and hurricanes also depends to a large extent on how well the basic services and infrastructure, the common goods of the community, stand up to the wind and rain accompanying these storms. Whereas individual families bear full responsibility for preparing their own shelter to withstand the effects of storms, they have a much more limited role in ensuring that their common services are safeguarded, yet one that cannot be neglected.

Non-governmental agencies involved in low income housing construction and upgrading have developed practical and low cost measures for increasing the resistance of self-built houses to hurricane force winds. Typical of efforts of this nature is the work performed by the Construction Resource and Development Centre

(CRDC) in Jamaica, which produced educational materials and organised workshops on house and roof reconstruction following Hurricane Gilbert.

The principal responsibility for introducing an awareness and concern in the community regarding the risk posed by hurricanes to the common good rests with the community leadership and local-or district-disaster coordinator, if such a function exists. It involves a lengthy process of identifying the issues, mobilising resources from within the community and from outside, and building support for common action.

Such a process consists of six steps: (1) making an inventory of lifeline networks and critical facilities; (2) learning the operation of these and their potential for disruption by a hurricane; (3) checking the vulnerability of the lifelines and critical facilities through field inspection and investigation; (4) establishing a positive working relationship with the agencies and companies that manage the infrastructure and services of the community; (5) developing an understanding of the total risk to the community; (6) formulating and implementing a mitigation strategy.

5.4.4.1. *Inventory of lifeline networks and critical facilities*

Lifeline networks and critical facilities are those elements in the economic and social infrastructure that provide essential goods and services to the population in towns and villages. Their proper functioning is a direct concern of the community, since disruption affects the entire population.

The community leadership should gradually build up an inventory of these elements by locating them in a first instance on a large-scale map (1:5,000 or 1:2,500) of the community. The base maps can be obtained from the town and country departments or physical planning offices. The road network should indicate the road hierarchy (highway, principal access to settlement, local streets) and the location of bridges and other civil works such as major road cuts and retaining walls. Similar treatment should be given to the electricity and telephone networks and the water system. Residential areas and areas of economic activity should also be identified.

Various sources can be tapped to obtain this information. Water, electricity, and telecommunication companies can be called upon to draw their networks on the maps for the area in question. The local representative of the ministry of public works or physical planning office can assist with the identification of the road network and the location of public facilities housing important services.

5.4.4.2. *Learning the operation of lifelines and facilities and their potential for disruption by Hurricane*

Community leaders should periodically organise brief sessions in which the engineers or managers responsible for the different lifelines and facilities can explain the workings of their systems to selected residents who may be involved in disaster preparedness and response. The maps that were prepared earlier should be helpful during these sessions, while at the same time particular details can be reviewed and updated. The focus of these sessions should be:

— Identification of the different elements that make up the system, their interaction, and their interdependency.

— How the different elements function, what can go wrong, and what the normal repair and maintenance procedures are.

— How each of the elements of the system can be affected by the forces accompanying a hurricane.

— What the consequences of a hurricane could be for the functioning of the system and for the users.

Lifeline networks

— Road network, with roads, bridges, road cuts and retaining walls, elevated roads, drainage works.

— Water system, with surface intakes, wells, pipelines, treatment plants, pumping stations, storage tanks or reservoirs, water mains, and distribution network.

— Electricity system, with generating plant, transmission lines, substations, transformers, and distribution network.

— Telecommunication, with ground station, exchanges, microwave transmission towers, aerial and underground cables, and open line distribution network.

— Sanitation system, with collector network, treatment plant and sewage fallout; public washrooms and toilet facilities; solid waste collection and disposal.

Critical facilities

— Hospitals, health centers, schools, churches.

— Fire stations, police stations, community centers, shelters, and other public buildings that house vital functions that play a role in emergencies.

Checking the vulnerability of the lifelines and facilities through field inspection and investigation

The vulnerability of buildings and infrastructure elements will be determined first of all by their location with respect to hazard-prone areas. Storm surges and wave action can inflict severe damage in waterfront and low-lying coastal areas; heavy rains accompanying the hurricanes can cause flash flooding or riverine flooding along the river banks and in low-lying areas; rain can also cause land slippages and mudslides on steep slopes and unstable roadcuts; and structures in exposed areas such as ridges and bluffs are particularly vulnerable to wind damage.

Hazard-prone areas should be systematically identified and located on the lifeline and critical facilities map, to show where lifeline networks and critical facilities may be especially vulnerable.

Establishing a Positive Working Relationship with the Agencies and Companies that Manage the Infrastructure and Services of the Community

Once the community leadership has collected a fair amount of information, a series of consultations should be organised with the engineers or managers responsible for each of the lifeline and critical facilities of the settlement, or with their local representatives, and further elaboration of the information collected thus far should take place.

Such consultations provide an opportunity for the community leadership to learn about the maintenance and emergency repair policies practiced in their settlements by the different agencies and utility companies, to get to know the officers responsible for carrying out emergency repairs, and to find out how to contact them under normal circumstances as well as in emergencies.

Good contacts between agency representatives and community leadership are of great help in exploring the coincidence of interest between the residents on the one hand and the service agencies and companies on the other. Through effectively managed participation by the residents in such tasks as monitoring the state of repair of the infrastructure or keeping drains clear, the community can receive better services at a lower cost to the agencies responsible. The actual hiring of workers or small firms from the settlement to execute some of the agencies' tasks should be encouraged wherever possible.

Formulating a Mitigation Strategy

The formulation of a strategy to introduce appropriate mitigation measures that respond to the community's priorities is the culmination of all the efforts expended on the vulnerability analysis and risk assessment.

It is important that the community leadership focus on identifying realistic mitigation measures and proposing a simple implementation strategy. The common pitfall of identifying measures that require substantial funding should be avoided by concentrating on non-structural measures.

Typical of the measures that should be emphasised are those that can be integrated into routine maintenance or upgrading of infrastructure; the avoidance of environmental degradation that can decrease the natural protective capacity of resources such as sand dunes, mangroves, and other natural vegetative coverage; and prevention by means of proper planning and design of new investments. It is also important to establish the role of the different governmental levels and agencies in the country in the implementation of a mitigation strategy.

6

Tornado

A tornado is a violently rotating column of air which is in contact with both a cumulonimbus cloud or, in rare cases, a cumulus cloud base and the surface of the earth. Tornadoes come in many sizes but are typically in the form of a visible condensation funnel, whose narrow end touches the earth and is often encircled by a cloud of debris.

Most tornadoes have wind speeds of 110 mph (177 km/h) or less, are approximately 250 feet (75 m) across, and travel a few miles (several kilometers) before dissipating. Some attain wind speeds of more than 300 mph (480 km/h), stretch more than a mile (1.6 km) across, and stay on the ground for dozens of miles (more than 100 km).

Although tornadoes have been observed on every continent except Antarctica, most occur in the United States. They also commonly occur in southern Canada, south-central and eastern Asia, east-central South America, Southern Africa, northwestern and central Europe, Italy, western and southeastern Australia, and New Zealand.

A tornado is not necessarily visible; however, the intense low pressure caused by the high wind speeds and rapid rotation (due to cyclostrophic balance) usually causes water vapor in the air to condense into a visible condensation funnel. The tornado is the vortex of wind, not the condensation cloud.

A funnel cloud is a visible condensation funnel with no associated strong winds at the surface. Not all funnel clouds evolve into a tornado. However, many tornadoes are preceded by a funnel cloud as the mesocyclonic rotation descends toward the ground. Most tornadoes produce strong winds at the surface while the visible funnel is still above the ground, so it is difficult to tell the difference between a funnel cloud and a tornado from a distance.

Occasionally, a single storm produces multiple tornadoes and mesocyclones. This process is known as cyclic tornadogenesis. Tornadoes produced from the same storm are referred to as a tornado family. Sometimes multiple tornadoes from distinct mesocyclones occur simultaneously.

Occasionally, several tornadoes are spawned from the same large-scale storm system. If there is no break in activity, this is considered a tornado outbreak, although there are various definitions. A period of several successive days with tornado outbreaks in the same general area (spawned by multiple weather systems) is a tornado outbreak sequence, occasionally called an *extended tornado outbreak.*

6.1. Types of Tornadoes

6.1.1. True Tornadoes

6.1.1.1. Multiple vortex tornado

A multiple vortex tornado is a type of tornado in which two or more columns of spinning air rotate around a common center. Multivortex structure can occur in almost any circulation, but is very often observed in intense tornadoes.

Satellite tornado

A satellite tornado is a term for a weaker tornado which forms very near a large, strong tornado contained within the same mesocyclone. The satellite tornado may appear to "orbit" the larger tornado, giving the appearance of one, large multi-vortex tornado. However, a satellite tornado is a distinct funnel, and is much smaller than the main funnel.

Waterspout

A waterspout is officially defined by the US National Weather Service simply as a tornado over water. However, researchers typically distinguish "fair weather" waterspouts from tornadic waterspouts.

Fair weather waterspouts are less severe but far more common, and are similar in dynamics to dust devils and landspouts. They form at the bases of cumulus congestus cloud towers in tropical and semitropical waters. They have relatively weak winds, smooth laminar walls, and typically travel very slowly, if at all.

Tornadic waterspouts are more literally "tornadoes over water". They can form over water like mesocyclonic tornadoes, or be a land tornado which crosses onto water. Since they form from severe thunderstorms and can be far more intense, faster, and longer-lived than fair weather waterspouts, they are considered far more dangerous.

Landspout

Landspout is an unofficial term for a tornado not associated with a mesocyclone. The name stems from their characterisation as essentially a "fair weather waterspout on land". Waterspouts and landspouts share many defining characteristics, including relative weakness, short lifespan, and a small, smooth condensation funnel which often does not reach the ground. Landspouts also create a distinctively laminar cloud of dust when they make contact with the ground, due to their differing mechanics from true mesoform tornadoes. Though usually weaker than classic tornadoes, they still produce strong winds and may cause serious damage.

6.1.1.2. Tornado-like circulations

Gustnado

A gustnado (gust front tornado) is a small, vertical swirl associated with a gust front or downburst. Because they are technically not associated with the cloud base, there is some debate as to whether or not gustnadoes are actually tornadoes. They are formed when fast moving cold, dry outflow air from a thunderstorm is blown through a mass of stationary, warm, moist air near the outflow boundary, resulting in a "rolling" effect (often exemplified through a roll cloud). If low level wind shear is strong enough, the rotation can be turned horizontally (or diagonally) and make contact with the ground. The result is a gustnado. They usually cause small areas of heavier rotational wind damage among areas of straight-line wind damage. It is also worth noting that since they are absent of any Coriolis influence from a mesocyclone, they seem to be alternately cyclonic and anticyclonic without preference.

Dust devil

A dust devil resembles a tornado in that it is a vertical swirling column of air. However, they form under clear skies and are rarely as strong as even the weakest tornadoes. They form when a strong convective updraft is formed near the ground on a hot day. If there is enough low level wind shear, the column of hot, rising air can develop a small cyclonic motion that can be seen near the ground. They are not considered tornadoes because they form during fair weather and are not associated with any actual cloud. However, they can, on occasion, result in major damage, especially in arid areas.

Winter Waterspout

A winter waterspout, also known as a snow devil or a snowspout, is an extremely rare meteorological phenomenon in which a vortex resembling that of a waterspout forms under the base of a snow squall.

Fire whirl

Tornado-like circulations occasionally occur near large, intense wildfires and are called fire whirls. They are not considered tornadoes except in the rare case where they connect to a pyrocumulus or other cumuliform cloud above. Fire whirls usually are not as strong as tornadoes associated with thunderstorms. However, they can produce significant damage.

Cold air vortex

A cold air vortex or *shear funnel* is a tiny, harmless funnel cloud which occasionally forms underneath or on the sides of normal cumuliform clouds, rarely causing any winds at ground-level. Their genesis and mechanics are poorly understood, as they are quite rare, short lived, and hard to spot (due to their non-rotational nature and small size).

6.1.2. Characteristics of Tornadoes

6.1.2.1. Shape

Most tornadoes take on the appearance of a narrow funnel, a few hundred yards (a few hundred meters) across, with a small cloud of debris near the ground. However, tornadoes can appear in many shapes and sizes.

Small, relatively weak landspouts may only be visible as a small swirl of dust on the ground. While the condensation funnel may not extend all the way to the ground, if associated surface winds are greater than 40 mph (64 km/h), the circulation is considered a tornado. Large single-vortex tornadoes can look like large wedges stuck into the ground, and so are known as *wedge tornadoes* or *wedges*. A wedge can be so wide that it appears to be a block of dark clouds, wider than the distance from the cloud base to the ground. Even experienced storm observers may not be able to tell the difference between a low-hanging cloud and a wedge tornado from a distance.

Tornadoes in the dissipating stage can resemble narrow tubes or ropes, and often curl or twist into complex shapes. These tornadoes are said to be *roping out*, or becoming a *rope tornado*. Multiple-vortex tornadoes can appear as a family of swirls circling a common center, or may be completely obscured by condensation, dust, and debris, appearing to be a single funnel.

In addition to these appearances, tornadoes may be obscured completely by rain or dust. These tornadoes are especially dangerous, as even experienced meteorologists might not spot them.

6.1.2.2. Size

In the United States, on average tornadoes are around 500

feet (150 m) across, and stay on the ground for 5 miles (8 km). Yet, there is an extremely wide range of tornado sizes, even for typical tornadoes. Weak tornadoes, or strong but dissipating tornadoes, can be exceedingly narrow, sometimes only a few feet across. A tornado was once reported to have a damage path only 7 feet (2 m) long. On the other end of the spectrum, wedge tornadoes can have a damage path a mile (1.6 km) wide or more. A tornado that affected Hallam, Nebraska on May 22, 2004 was at one point 2.5 miles (4 km) wide at the ground.

In terms of path length, the Tri-State Tornado, which affected parts of Missouri, Illinois, and Indiana on March 18, 1925, was officially on the ground continuously for 219 miles (352 km). Many tornadoes which appear to have path lengths of 100 miles or longer are actually a family of tornadoes which have formed in quick succession; however, there is no substantial evidence that this occurred in the case of the Tri-State Tornado. In fact, modern reanalysis of the path suggests that the tornado began 15 miles (24 km) further west than previously thought.

6.1.2.3. Appearance

Tornadoes can have a wide range of colors, depending on the environment in which they form. Those which form in a dry environment can be nearly invisible, marked only by swirling debris at the base of the funnel. Condensation funnels which pick up little or no debris can be gray to white. While travelling over a body of water as a waterspout, they can turn very white or even blue. Funnels which move slowly, ingesting a lot of debris and dirt, are usually darker, taking on the color of debris. Tornadoes in the Great Plains can turn red because of the reddish tint of the soil, and tornadoes in mountainous areas can travel over snow-covered ground, turning brilliantly white.

Lighting conditions are a major factor in the appearance of a tornado. A tornado which is "back-lit" (viewed with the sun behind it) appears very dark. The same tornado, viewed with the sun at the observer's back, may appear gray or brilliant white. Tornadoes which occur near the time of sunset can be many different colors, appearing in hues of yellow, orange, and pink.

Dust kicked up by the winds of the parent thunderstorm, heavy rain and hail, and the darkness of night are all factors which can reduce the visibility of tornadoes. Tornadoes occurring in these conditions are especially dangerous, since only radar observations, or possibly the sound of an approaching tornado, serve as any warning to those in the storm's path. Fortunately most significant tornadoes form under the storm's *rain-free base,* or the area under the thunderstorm's updraft, where there is little or no rain. In addition, most tornadoes occur in the late afternoon, when the bright sun can penetrate even the thickest clouds. Also, night-time tornadoes are often illuminated by frequent lightning.

There is mounting evidence, including Doppler On Wheels mobile radar images and eyewitness accounts, that most tornadoes have a clear, calm center with extremely low pressure, akin to the eye of tropical cyclones. This area would be clear (possibly full of dust), have relatively light winds, and be very dark, since the light would be blocked by swirling debris on the outside of the tornado. Lightning is said to be the source of illumination for those who claim to have seen the interior of a tornado.

6.1.2.4. Rotation

Tornadoes normally rotate cyclonically in direction (counterclockwise in the northern hemisphere, clockwise in the southern). While large-scale storms always rotate cyclonically due to the Coriolis effect, thunderstorms and tornadoes are so small that the direct influence of Coriolis effect is inconsequential, as indicated by their large Rossby numbers. Supercells and tornadoes rotate cyclonically in numerical simulations even when the Coriolis effect is neglected. Low-level mesocyclones and tornadoes owe their rotation to complex processes within the supercell and ambient environment.

Approximately 1% of tornadoes rotate in an anticyclonic direction. Typically, only landspouts and gustnados rotate anticyclonically, and usually only those which form on the anticyclonic shear side of the descending rear flank downdraft in a cyclonic supercell. However, on rare occasions, anticyclonic tornadoes form in association with the mesoanticyclone of an anticyclonic supercell, in the same manner as the typical cyclonic tornado, or as a companion tornado—either as a satellite tornado or associated with anticyclonic eddies within a supercell.

6.1.2.5. Sound and seismology

Tornadoes emit widely on the acoustics spectrum and the sounds are cased by multiple mechanisms. Various sounds of tornadoes have been reported throughout time, mostly related to familiar sounds for the witness and generally some variation of a whooshing roar. Popularly reported sounds include a freight train, rushing rapids or waterfall, a jet engine from close proximity, or combinations of these. Many tornadoes are not audible from much distance; the nature and propagation distance of the audible sound depends on atmospheric conditions and topography.

The winds of the tornado vortex and of constituent turbulent eddies, as well as airflow interaction with the surface and debris, contribute to the sounds. Funnel clouds also produce sounds. Funnel clouds and small tornadoes are reported as whistling, whining, humming, or the buzzing of innumerable bees or electricity, or more or less harmonic, whereas many tornadoes are reported as a continuous, deep rumbling, or an irregular sound of "noise".

Since many tornadoes are audible only in very close proximity, sound is not reliable warning of a tornado. And, any strong, damaging wind, even a severe hail volley or continuous thunder in a thunderstorm may produce a roaring sound. Tornadoes also produce identifiable inaudible infrasonic signatures. Unlike audible signatures, tornadic signatures have been isolated; due to the long distance propagation of low-frequency sound, efforts are ongoing to develop tornado prediction and detection devices with additional value in understanding tornado morphology, dynamics, and creation. Tornadoes also produce a detectable seismic signature, and research continues on isolating it and understanding the process.

6.1.2.6. Electromagnetic, lightning, and other effects

Tornadoes emit on the electromagnetic spectrum, for example, with sferics and E-field effects detected. The effects vary, mostly with little observed consistency. Correlations with patterns of lightning activity have also been observed, but little in way of consistent correlations have been advanced. Tornadic storms do not contain more lightning than other storms, and some tornadic cells never contain lightning. More often that not, overall cloud-to-ground (CG) lightning activity decreases as a tornado reaches the surface and returns to the baseline level when the tornado lifts. In many cases, very intense tornadoes and thunderstorms exhibit an increased and anomalous dominance in positive polarity CG discharges. Electromagnetics and lightning have little to nothing to do directly with what drives tornadoes (tornadoes are basically a thermodynamic phenomenon), though there are likely connections with the storm and environment affecting both phenomena.

Luminosity has been reported in the past, and is probably due to misidentification of external light sources such as lightning, city lights, and power flashes from broken lines, as internal sources are now uncommonly reported and are not known to ever been recorded.

In addition to winds, tornadoes also exhibit changes in atmospheric variables such as temperature, moisture, and pressure. For example, on June 24, 2003 near Manchester, South Dakota, a probe measured a 100 mbar (hPa) (2.95 inHg) pressure deficit. The pressure dropped gradually as the vortex approached then dropped extremely rapidly to 850 mbar (hPa) (25.10 inHg) in the core of the violent tornado before rising rapidly as the vortex moved away, resulting in a V-shape pressure trace. Temperature tends to decrease and moisture content to increase in the immediate vicinity of a tornado.

6.1.2.7. Life cycle

Supercell relationship

Tornadoes often develop from a class of thunderstorms known as supercells. Supercells contain mesocyclones, an area of organised rotation a few miles up in the atmosphere, usually 1–6 miles (2–10 km) across. Most intense tornadoes (EF3 to EF5 on the Enhanced Fujita Scale) develop from supercells. In addition to tornadoes, very heavy rain, frequent lightning, strong wind gusts, and hail are common in such storms.

Most tornadoes from supercells follow a recognisable life cycle. That begins when increasing rainfall drags with it an area of quickly descending air known as the rear flank downdraft (RFD). This downdraft accelerates as it approaches the ground, and drags the supercell's rotating mesocyclone towards the ground with it.

Formation

As the mesocyclone approaches the ground, a visible condensation funnel appears to descend from the base of the storm, often from a rotating wall cloud. As the funnel descends, the RFD also reaches the ground, creating a gust front that can cause damage a good distance from the tornado. Usually, the funnel cloud becomes a tornado within minutes of the RFD reaching the ground.

Maturity

Initially, the tornado has a good source of warm, moist inflow to power it, so it grows until it reaches the *mature stage*. This can last anywhere from a few minutes to more than an hour, and during it a tornado often causes the most damage, and in rare cases can be more than one mile across. Meanwhile, the RFD, now an area of cool surface winds, begins to wrap around the tornado, cutting off the inflow of warm air which feeds the tornado.

Demise

As the RFD completely wraps around and chokes off the tornado's air supply, the vortex begins to weaken, and become thin and rope-like. This is the *dissipating stage*; often lasting no more than a few minutes, after which the tornado fizzles. During this stage the shape of the tornado becomes highly influenced by the winds of the parent storm, and can be blown into fantastic patterns.

As the tornado enters the dissipating stage, its associated mesocyclone often weakens as well, as the rear flank downdraft cuts off the inflow powering it. In particularly intense supercells tornadoes can develop cyclically. As the first mesocyclone and associated tornado dissipate, the storm's inflow may be concentrated into a new area closer to the center of the storm. If a new mesocyclone develops, the cycle may start again, producing one or more new tornadoes. Occasionally, the old (*occluded*) mesocyclone and the new mesocyclone produce a tornado at the same time.

Though this is a widely-accepted theory for how most tornadoes form, live, and die, it does not explain the formation of smaller tornadoes, such as landspouts, long-lived tornadoes, or tornadoes with multiple vortices. These each have different mechanisms which influence their development—however, most tornadoes follow a pattern similar to this one.

Intensity and damage

The Fujita scale and the Enhanced Fujita Scale rate tornadoes by damage caused. The Enhanced Fujita Scale was an upgrade to the older Fujita scale, with engineered (by expert elicitation) wind estimates and better damage descriptions, but was designed so that a tornado rated on the Fujita scale would receive the same numerical rating. An EF0 tornado will likely damage trees but not substantial structures, whereas an EF5 tornado can rip buildings off their foundations leaving them bare and even deform large skyscrapers. The similar TORRO scale ranges from a T0 for extremely weak tornadoes to T11 for the most powerful known tornadoes. Radar data, photogrammetry, and ground swirl patterns (cycloidal marks) may also be analysed to determine intensity and award a rating.

Tornadoes vary in intensity regardless of shape, size, and location, though strong tornadoes are typically larger than weak tornadoes. The association with track length and duration also varies, although longer track tornadoes tend to be stronger. In the case of violent tornadoes, only a small portion of the path is of violent intensity, most of the higher intensity from subvortices.

In the United States, 80% of tornadoes are EF0 and EF1 (T0 through T3) tornadoes. The rate of occurrence drops off quickly with increasing strength—less than 1% are violent tornadoes, stronger than EF4, T8.

Outside the United States, areas in south-central Asia, and perhaps portions of southeastern South America and southern Africa, violent tornadoes are extremely rare. This is apparently mostly due to the lesser number of tornadoes overall, as research shows that tornado intensity distributions are fairly similar worldwide. A few significant tornadoes occur annually in Europe, Asia, southern Africa, and southeastern South America, respectively.

The United States has the most tornadoes of any country, about four times more than estimated in all of Europe, not including waterspouts. This is mostly due to the unique geography of the continent. North America is a relatively large continent that extends from the tropical south into arctic areas, and has no major east-west mountain range to block air flow between these two areas. In the middle latitudes, where most tornadoes of the world occur, the Rocky Mountains block moisture and atmospheric flow, allowing drier air at mid-levels of the troposphere, and causing cyclogenesis downstream to the east of the mountains. The desert Southwest also feeds drier air and the dry line, while the Gulf of Mexico fuels abundant low-level moisture.

This unique topography allows for many collisions of warm and cold air, the conditions that breed strong, long-lived storms many times a year. A large portion of these tornadoes form in an area of the central United States known as Tornado Alley. This area extends into Canada, particularly Ontario and the Prairie Provinces. Strong tornadoes also occasionally occur in northern Mexico.

The United States averages about 1,200 tornadoes per year. The Netherlands has the highest average number of recorded tornadoes per area of any country (more than 20, or 0.0013 per sq mi (0.00048 per km^2), annually), followed by the UK (around 33, or 0.00035 per sq mi (0.00013 per km^2), per year), but most are small and cause minor damage. In absolute number of events, ignoring area, the UK experiences more tornadoes than any other European country, excluding waterspouts.

Bangladesh and surrounding areas of eastern India suffer from tornadoes of equal severity to those in the US with more regularity than any other region in the world, but these tend to be under-reported due to the scarcity of media coverage in third-world countries. They kill about 179 people per year in Bangladesh, much more than in the US. This is likely due to the density of population, poor quality of construction, lack of tornado safety knowledge, and other factors. Other areas of the world that have frequent tornadoes include South Africa, parts of Argentina, Paraguay, and southern Brazil, as well as portions of Europe, Australia and New Zealand, and far eastern Asia.

Tornadoes are most common in spring and least common in winter. Since autumn and spring are transitional periods (warm to cool and vice versa) there are more chances of cooler air meeting with warmer air, resulting in thunderstorms. Tornadoes can also be caused by landfalling tropical cyclones, which tend to occur in the late summer and autumn. But favorable conditions can occur at any time of the year.

Tornado occurrence is highly dependent on the time of day, because of solar heating. Worldwide, most tornadoes occur in the late afternoon, between 3 and 7 pm local time, with a peak near 5 pm. However, destructive tornadoes can occur at any time of day. The Gainesville Tornado of 1936, one of the deadliest tornadoes in history, occurred at 8:30 am local time.

6.1.2.8. Weather forecasting

Weather forecasting is handled regionally by many national and international agencies. For the most part, they are also in charge of the prediction of conditions conducive to tornado development.

Severe thunderstorm warnings are provided to Australia by the Bureau of Meteorology. The country is in the middle of an upgrade to Doppler radar systems, with their first benchmark of installing six new radars reached in July 2006.

The European Union founded a project in 2002 called the European Severe Storms virtual Laboratory, or ESSL, which is meant to fully document tornado occurrence across the continent. The ESTOFEX (European Storm Forecast Experiment) arm of the project also issues one day forecasts for severe weather likelihood. In Germany, Austria, and Switzerland, an organisation known as TorDACH collects information regarding tornadoes, waterspouts, and downbursts from Germany, Austria, and Switzerland. A secondary goal is collect all severe weather information. This project is meant to fully document severe weather activity in these three countries.

In the United Kingdom, the Tornado and Storm Research Organisation (TORRO) makes experimental predictions. The Met Office provides official forecasts for the UK.

In the United States, generalised severe weather predictions are issued by the Storm Prediction Center, based in Norman, Oklahoma. For the next one, two, and three days, respectively, they will issue categorical and probabilistic forecasts of severe weather, including tornadoes. There is also a more general forecast issued for the four to eight day period. Just prior to the expected onset of an organised severe weather threat, SPC issues severe thunderstorm and tornado watches, in collaboration with local National Weather Service offices. Warnings are issued by local National Weather Service offices when a severe thunderstorm or tornado is occurring or imminent.

In Japan, predictions and study of tornadoes in Japan are handled by the Japan Meteorological Agency. In Canada, weather forecasts and warnings, including tornadoes, are produced by the Meteorological Service of Canada, a division of Environment Canada.

Rigorous attempts to warn of tornadoes began in the United States in the mid-20th century. Before the 1950s, the only method of detecting a tornado was by someone seeing it on the ground. Often, news of a tornado would reach a local weather office after the storm.

But, with the advent of weather radar, areas near a local office could get advance warning of severe weather. The first public tornado warnings were issued in 1950 and the first tornado watches and convective outlooks in 1952. In 1953 it was confirmed that hook echoes are associated with tornadoes. By recognising these radar signatures, meteorologists could detect thunderstorms likely producing tornadoes from dozens of miles away.

Storm spotting

In the mid 1970s, the US National Weather Service (NWS) increased its efforts to train storm spotters to spot key features of storms which indicate severe hail, damaging winds, and tornadoes, as well as damage itself and flash flooding. The program was called Skywarn, and the spotters were local sheriff's deputies, state troopers, firefighters, ambulance drivers, amateur radio operators, civil defense (now emergency management) spotters, storm chasers, and ordinary citizens. When severe weather is anticipated, local weather service offices request that these spotters look out for severe weather, and report any tornadoes immediately, so that the office can issue a timely warning.

Usually spotters are trained by the NWS on behalf of their respective organisations, and report to them. The organisations activate public warning systems such as sirens and the Emergency Alert System, and forward the report to the NWS. There are more than 230,000 trained Skywarn weather spotters across the United States.

In Canada, a similar network of volunteer weather watchers, called Canwarn, helps spot severe weather, with more than 1,000 volunteers. In Europe, several nations are organising spotter networks under the auspices of Skywarn Europe and the Tornado and Storm Research Organisation (TORRO) has maintained a network of spotters in the United Kingdom since the 1970s.

Storm spotters are needed because radar systems such as NEXRAD do not detect a tornado; only indications of one. Radar may give a warning before there is any visual evidence of a tornado or imminent tornado, but ground truth from an observer can either verify the threat or determine that a tornado is not imminent. The spotter's ability to see what radar cannot is especially important as distance from the radar site increases, because the radar beam becomes progressively higher in altitude further away from the radar, chiefly due to curvature of Earth, and the beam also spreads out. Therefore, when far from a radar, only high in the storm is observed and the important areas are not sampled, and data resolution also suffers. Also, some meteorological situations leading to tornadogenesis are not readily detectable by radar and on occasion tornado development may occur more quickly than radar can complete a scan and send the batch of data.

Storm spotters are trained to discern whether a storm seen from a distance is a supercell. They typically look to its rear, the main region of updraft and inflow. Under the updraft is a rain-free base, and the next step of tornadogenesis is the formation of a rotating wall cloud. The vast majority of intense tornadoes occur with a wall cloud on the backside of a supercell.

Evidence of a supercell comes from the storm's shape and structure, and cloud tower features such as a hard and vigorous updraft tower, a persistent, large overshooting top, a hard anvil (especially when backsheared against strong upper level winds), and a corkscrew look or striations. Under the storm and closer to where most tornadoes are found, evidence of a supercell and likelihood of a tornado includes inflow bands (particularly when curved) such as a "beaver tail", and other clues such as strength of inflow, warmth and moistness of inflow air, how outflow- or inflow-dominant a storm appears, and how far is the front flank precipitation core from the wall cloud. Tornadogenesis is most likely at the interface of the updraft and front flank downdraft, and requires a balance between the outflow and inflow.

Only wall clouds that rotate spawn tornadoes, and usually precede the tornado by five to thirty minutes. Rotating wall clouds are the visual manifestation of a mesocyclone. Barring a low-level boundary, tornadogenesis is highly unlikely unless a rear flank downdraft occurs, which is usually visibly evidenced by evaporation of cloud adjacent to a corner of a wall cloud.

A tornado often occurs as this happens or shortly after; first, a funnel cloud dips and in nearly all cases by the time it reaches halfway down, a surface swirl has already developed, signifying a tornado is on the ground before condensation connects the surface circulation to the storm. Tornadoes may also occur without wall clouds, under flanking lines, and on the leading edge. Spotters watch all areas of a storm, and the cloud base and surface.

Radar

Today, most developed countries have a network of weather radars, which remains the main method of detecting signatures likely associated with tornadoes. In the United States and a few other countries, Doppler radar stations are used. These devices measure the velocity and radial direction (towards or away from the radar) of the winds in a storm, and so can spot evidence of rotation in storms from more than a hundred miles away.

Also, most populated areas on Earth are now visible from the Geostationary Operational Environmental Satellites (GOES), which aid in the nowcasting of tornadic storms.

6.1.3. Extreme Tornado

The most extreme tornado in recorded history was the Tri-State Tornado which roared through parts of Missouri, Illinois, and Indiana on March 18, 1925. It was likely an F5, though tornadoes were not ranked on any scale in that era. It holds records for longest path length (219 miles, 352 km), longest duration (about 3.5 hours), and fastest forward speed for a significant tornado (73 mph, 117 km/h) anywhere on earth. In addition, it is the deadliest single tornado in United States history (695 dead). It was also the second costliest tornado in history at the time, but has been surpassed by several others non-normalised. When costs are normalised for wealth and inflation, it still ranks third today.

The deadliest tornado in world history was the Daultipur-Salturia Tornado in Bangladesh on April 26, 1989, killing approximately 1300 people.

The most extensive tornado outbreak on record, in almost every category, was the Super Outbreak, which affected a large area of the central United States and extreme southern Ontario in Canada on April 3 and April 4, 1974. Not only did this outbreak feature an incredible 148 tornadoes in only 18 hours, but an unprecedented number of them were violent; six were of F5 intensity, and twenty-four F4. This outbreak had a staggering *sixteen* tornadoes on the ground at the same time at the peak of the outbreak. More than 300 people, possibly as many as 330, were killed by tornadoes during this outbreak.

While it is nearly impossible to directly measure the most violent tornado wind speeds (conventional anemometers would be destroyed by the intense winds), some tornadoes have been scanned by mobile Doppler radar units, which can provide a good estimate of the tornado's winds.

The highest wind speed ever measured in a tornado, which is also the highest wind speed ever recorded on the planet, is 301 ± 20 mph (484 ± 32 km/h) in the F5 Moore, Oklahoma tornado. Though the reading was taken about 100 feet (30 m) above the ground, this is a testament to the power of the strongest tornadoes.

Storms which produce tornadoes can feature intense updrafts (sometimes exceeding 150 mph, 240 km/h). Debris from a tornado can be lofted into the parent storm and carried a very long distance. A tornado which affected Great Bend, Kansas in November, 1915 was an extreme case, where a "rain of debris" occurred 80 miles (130 km) from the town, a sack of flour was found 110 miles (177 km) away, and a cancelled check from the Great Bend bank was found in a field outside of Palmyra, Nebraska, 305 miles (491 km) to the northeast.

6.2. Precautionary Measures

Though tornadoes can strike in an instant, there are precautions and preventative measures that people can take to increase the chances of surviving a tornado. Authorities such as the Storm Prediction Center advise having a tornado plan. When a tornado warning is issued, going to a basement or an interior first-floor room of a sturdy building greatly increases chances of survival. In tornado-prone areas, many buildings have storm cellars on the property. These underground refuges have saved thousands of lives.

Some countries have meteorological agencies which distribute tornado forecasts and increase levels of alert of a possible tornado (such as tornado watches and warnings in the United States and Canada). Weather radios provide an alarm when a severe weather advisory is issued for the local area, though these are mainly available only in the United States.

Unless the tornado is far away and highly visible, meteorologists advise that drivers park their vehicles far to the side of the road (so as not to block emergency traffic), and find a sturdy shelter. If no sturdy shelter is nearby, getting low in a ditch is the next best option. Highway overpasses are extremely bad shelter during tornadoes.

Myths and Misconceptions

One of the most persistent myths associated with tornadoes is that opening windows will lessen the damage caused by the tornado. While there is a large drop in atmospheric pressure inside a strong tornado, it is unlikely that the pressure drop would be enough to cause the house to explode. Some research indicates that opening windows may actually increase the severity of the tornado's damage. Regardless of the validity of the explosion claim, time would be better spent seeking shelter before a tornado than opening windows. A violent tornado can destroy a house whether its windows are open or closed.

Another commonly held belief is that highway overpasses provide adequate shelter from tornadoes. On the contrary, a highway overpass is a dangerous place during a tornado. In the Oklahoma Tornado Outbreak of May 3, 1999, three highway overpasses were directly struck by tornadoes, and at all three locations there was a fatality, along with many life-threatening injuries. The small area under the overpasses created a kind of wind tunnel, increasing the wind's speed, making the situation worse. By comparison, during the same tornado outbreak, more than 2000 homes were completely destroyed, with another 7000 damaged, and yet only a few dozen people died in their homes.

An old belief is that the southwest corner of a basement provides the most protection during a tornado. The safest place is the side or corner of an underground room opposite the tornado's direction of approach (usually the northeast corner), or the central-most room on the lowest floor. Taking shelter under a sturdy table, in a basement, or under a staircase increases chances of survival even more.

Finally, there are areas which people believe to be protected from tornadoes, whether by a major river, a hill or mountain, or even protected by "spirits". Tornadoes have been known to cross major rivers, climb mountains, and affect valleys. As a general rule, no area is "safe" from tornadoes, though some areas are more susceptible than others.

7

Man-made Disasters

Man-made Disasters disasters are events which are caused by man, either intentionally or by accident, which that can directly or indirectly cause severe threats, either directly or indirectly, to public health and/or well-being. Because their occurrence is unpredictable, man-made disasters pose an especially challenging threat which that must be dealt with through vigilance, and proper preparedness and response. Information on the major sources of man-made disasters have beenare provided here to help educate the public on their cause and effects as they relate to emergency planning.

7.1. Bioterrorism

A bioterrorism attack is the deliberate release of viruses, bacteria, or other germs (agents) used to cause illness or death in people, animals, or plants. These agents are typically found in nature, but it is possible that they could be changed to increase their ability to cause disease, make them resistant to current medicines, or to increase their ability to be spread into the environment. Biological agents can be spread through the air, through water, or in food. Terrorists may use biological agents because they can be extremely difficult to detect and do not cause illness for several hours to several days. Some bioterrorism agents, like the smallpox virus, can be spread from person to person and some, like anthrax, can not.

7.1.1. Biological Agents

Bioterrorism agents can be separated into three categories, depending on how easily they can be spread and the severity of illness or death they cause. Category A agents are considered the highest risk and Category C agents are those that are considered emerging threats for disease.

7.1.1.1. Category A agents

These are biological agents with both a high potential for adverse public health impact and that also have a serious potential for large-scale dissemination. The Category A agents are anthrax, smallpox, plague, botulism, tularemia, and viral hemorrhagic fevers.

Anthrax

Anthrax is a non-contagious disease. An anthrax vaccine does exist but requires many injections and has side effects that render it unsuitable for general use.

Smallpox

Smallpox is a highly contagious virus. It transmits easily through the atmosphere and has a high mortality rate (20-40%). Smallpox was eliminated in the world in the 1970s, thanks to a worldwide vaccination program. However, some virus samples are still available in Russian and American laboratories. Some believe that after the collapse of the Soviet Union, cultures of smallpox have become available in other countries. Although people born pre-1970 will have been vaccinated for smallpox under the WHO program, the effectiveness of vaccination is limited since the vaccine provides high level of immunity for only 3 to 5 years. As a biological weapon smallpox is dangerous because of the highly contagious nature of both the infected and their pox. Smallpox occurs only in humans, and has no external hosts or vectors.

Botulinum toxin

Botulinum toxin is one of the deadliest toxins known, and is produced by the bacterium Clostridium botulinum. Botulism causes death by respiratory failure and paralysis. It is also easy to obtain since it is found in the cosmetic products Botox and Dysport.

Ebola

Ebola is a viral hemorrhagic fever, with fatality rates ranging from 50-90%. No cure currently exists, although vaccines are in development. The United States and the erstwhile Soviet Union both investigated the use of ebola for biological warfare, and the Aum Shinrikyo group possessed cultures of the virus. Ebola kills its victims through multiple organ failure and hypovolemic shock.

Plague

Plague is a disease caused by the Yersinia pestis bacterium. Rodents are the normal host of plague, and the disease is transmitted to humans by flea bites and occasionally by aerosol in the form of pneumonic plague. The disease has a history of use in biological warfare dating back many centuries, and is considered a threat due to its ease of culture and ability to remain in circulation among local rodents for a long period of time.

Marburg

Marburg is a viral hemorrhagic fever virus first discovered in Marburg, Germany. Fatality rates range from 25-100%, and although a vaccine is in development, no treatments currently exist aside from supportive care. As with ebola, basic barrier nursing significantly reduces the virulence of the virus.

Tularemia

Tularemia, or rabbit fever, is a generally non-lethal and severely incapacitating disease caused by the Francisella tularensis bacterium. It has been widely produced for biological warfare due to its highly infective nature, and ease of aerosolisation.

7.1.1.2. Category B agents

Category B agents are moderately easy to disseminate and have low mortality rates.

— Brucellosis (*Brucella* species) Brucellosis is an infectious disease caused by the bacteria of the genus Brucella. These bacteria are primarily passed among animals, and they cause disease in many different vertebrates. Various Brucella species affect sheep, goats, cattle, deer, elk, pigs, dogs, and several other animals. Humans become infected by coming in contact with animals or animal products that are contaminated with these bacteria. In humans brucellosis can cause a range of symptoms that are similar to the flu and may include fever, sweats, headaches, back pains, and physical weakness. Severe infections of the central nervous systems or lining of the heart may occur. Brucellosis can also cause long-lasting or chronic symptoms that include recurrent fevers, joint pain, and fatigue

— Epsilon toxin of Clostridium perfringens
— Food safety threats (e.g., Salmonella *species,* E coli O157:H7, Shigella, Stash)
  — Glanders (*Burkholderia mallei*)
  — Melioidosis (Burkholderia *pseudomallei*)
  — Psittacosis (*Chlamydia psittaci*)
  — Q fever (*Coxiella burnetii*)
  — Ricin toxin from *Ricinus communis* (castor beans)
  — Staphylococcal enterotoxin B
  — Typhus (*Rickettsia prowazekii*)
  — Viral encephalitis (alphaviruses, e.g.: Venezuelan equine encephalitis, eastern equine encephalitis, western equine encephalitis)
  — Water supply threats (e.g., Vibrio cholerae, Cryptosporidium *parvum,* Cholera)

7.1.1.3. Category C agents

Category C agents are pathogens that might be engineered for mass dissemination because they are easy to produce and have potential for high morbidity or mortality (examples: nipah virus, hantavirus and multi-drug resistant Tuberculosis (MTB).

7.1.2. Modern Bioterrorist Incidents

7.1.2.1. 1915-16 livestock sabotage by Germany

Dr Anton Dilger, a German-American physician, worked for Germany in the U.S. (Chevy Chase and Baltimore) in 1915 and 1916 with cultures of anthrax and glanders with the intention of biological sabotage on behalf of the German government. Other German agents are known to have undertaken similar sabotage efforts during WWI in Norway, Spain, Romania and Argentina.

7.1.2.2. 1984 Rajneeshee Salmonella attack

In 1984, followers of the Bhagwan Shree Rajneesh attempted to control a local election by incapacitating the local population by infecting salad bars in eleven restaurants, doorknobs, produce in grocery stores and other public domains with Salmonella typhimurium in the city of The Dalles, Oregon. The attack caused about 751 people to get sick (no fatalities). This incident was the first known bioterrorist attack in the United States in the 20th century.

7.1.2.3. 2001 anthrax attack

In September and October of 2001, several cases of anthrax broke out in the United States in the 2001 anthrax attacks, caused deliberately. This was a well-publicised act of bioterrorism. It motivated efforts to define biodefense and biosecurity, where more limited definitions of biosafety had focused on unintentional or accidental impacts of agricultural and medical technologies.

7.1.3. Defending Strategies

Planning may involve the development of biological identification systems. Until recently in the United States of America, most biological defense strategies have been geared to protecting soldiers on the battlefield rather than ordinary people in cities. Financial cutbacks have limited the tracking of disease outbreaks. Some outbreaks, such as food poisoning due to *E. coli* or *Salmonella,* could be of either natural or deliberate origin.

7.1.3.1. Biosurveillance strategies

In 1999, the University of Pittsburgh's Center for Biomedical Informatics deployed the first automated bioterrorism detection system, called RODS (Real-Time Outbreak Disease Surveillance). RODS is designed to draw collect data from many data sources and use them to perform signal detection, that is, to detect the a possible bioterrorism event at the earliest possible moment. RODS, another systems like it, collect data from sources including clinic data, laboratory data, and data from over-the-counter drug sales. In 2000, Michael Wagner, the codirector of the RODS laboratory, and Ron Aryel, a subcontractor, conceived of the idea of obtaining live data feeds from "non-traditional" (non-health-care) data sources. The RODS laboratory's first efforts eventually led to the establishment of the National Retail Data Monitor, a system which collects data from 20,000 retail locations nation-wide.

On February 5, 2002, President Bush visited the RODS laboratory and used it as a model for a $300 million spending proposal to equip all 50 states with biosurveillance systems. In a speech delivered at the nearby Masonic temple, Bush compared the RODS system to a modern "DEW" line.

The principles and practices of biosurveillance, a new interdisciplinary science, were defined and described in Handbook of Biosurveillance, edited by Michael Wagner, Andrew Moore and Ron Aryel, and published in 2006 by Elsevier's Academic Press division. Biosurveillance is the science of real-time disease outbreak detection. Its principles apply to both natural and man-made epidemics (bioterrorism).

Data which potentially could assist in early detection of a bioterrorism event include many categories of information. Health-related data such as that from hospital computer systems, clinical laboratories, electronic health record systems, medical examiner record-keeping systems, 911 call center computers, and veterinary medical record systems could be of help; researchers are also considering the utility of data generated by ranching and feedlot operations, food processors, drinking water systems, school attendance recording, and physiologic monitors, among others. Intuitively, one would expect systems which collect more than one type of data to be more useful than systems which collect only one type of information (such as single-purpose laboratory or 911 call-center based systems), and be less prone to false alarms, and this appears to be the case.

In Europe, disease surveillance is beginning to be organised on the continent-wide scale needed to track a biological emergency. The system not only monitors infected persons, but attempts to discern the origin of the outbreak.

Researchers are experimenting with devices to detect the existence of a threat:

— tiny electronic chips that would contain living nerve cells to warn of the presence of bacterial toxins (identification of broad range toxins)

— fiber-optic tubes lined with antibodies coupled to light-emitting molecules (identification of specific pathogens, such as anthrax, botulinum, ricin)

7.1.4. Limitations of Bioterrorism

Bioterrorism is inherently limited as a warfare tactic because of the uncontrollable nature of the agent involved. A biological weapon is useful to a terrorist group mainly as a method of creating mass panic and disruption to a society. However, technologists such as Bill Joy have warned of the potential power which genetic engineering might place in the hands of future bio-terrorists; a bacterial agent might be engineered for genetic or geographical selectivity.

The genomic revolution requires scientists to follow a recognised Code of Conduct. The 'dual-use' technology dilemma implicates issues further; good scientific inventions can be reapplied along a sinister vector.

7.2. Chemical Agents and Hazardous Materials

Chemicals are found everywhere. They purify drinking water, increase crop production, and simplify household chores. But chemicals also can be hazardous to humans or the environment if used or released improperly. Hazards can occur during production, storage, transportation, use, or disposal. You and your community are at risk if a chemical is used unsafely or released in harmful amounts into the environment where you live, work, or play.

Hazardous materials in various forms can cause death, serious injury, long-lasting health effects, and damage to buildings, homes, and other property. Many products containing hazardous chemicals are used and stored in homes routinely. These products are also shipped daily on the nation's highways, railroads, waterways, and pipelines.

Chemical manufacturers are one source of hazardous materials, but there are many others, including service stations, hospitals, and hazardous materials waste sites.

Varying quantities of hazardous materials are manufactured, used, or stored at an estimated 4.5 million facilities in the United States—from major industrial plants to local dry cleaning establishments or gardening supply stores.

Hazardous materials come in the form of explosives, flammable and combustible substances, poisons, and radioactive materials. These substances are most often released as a result of transportation accidents or because of chemical accidents in plants. Examples of some hazardous materials include:

— paints

— drugs

— cosmetics
— cleaning chemicals
— degreasers
— detergents
— gas cylinders
— refrigerant gases
— pesticides
— herbicides
— diesel fuel
— petrol
— liquefied petroleum gas
— welding fume.

Hazardous materials can cause adverse health effects such as asthma, skin rashes, allergic reactions, allergic sensitisation, cancer, and other long term diseases from exposure to substances.

7.2.1. Material Safety Data Sheets (MSDS)

An MSDS is a document containing important information about a hazardous substance and must state:

— a hazardous substance's product name
— the chemical and generic name of certain ingredients
— the chemical and physical properties of the hazardous substance
— health hazard information
— precautions for safe use and handling
— the manufacturer's or importer's name, Australian address and telephone number.

The MSDS provides employers, self-employed persons, workers and other health and safety representatives with the necessary information to safely manage the risk from hazardous substance exposure.

It is important that everyone in the workplace knows how to read and interpret a MSDS.

7.2.1.1. Access to MSDS

Access to a MSDS can be provided in several ways including:

— paper and microfiche copy collections of MSDS with microfiche readers open to use by all workers

— computerised and internet MSDS databases.

The register of MSDS should be used as an information tool to make sure everyone is involved in managing hazardous substances exposure at the workplace. A MSDS should be reviewed whenever there is:

— a change in formulation which:

— affects the hazardous properties of the substance

— alters the form, appearance or mode of application of the substance

— a change to the hazardous substance which alters its health and/or safety hazard or risk

— new health and/or safety information on the hazardous substance such as exposure standard changes or a substance previously considered not harmful is now established to be harmful (e.g. carcinogenic

— at least every five years.

In respect of MSDS and labels, employers and self-employed persons must:

— Obtain an MSDS of a hazardous substance from the supplier.

— Keep a register containing a list of all hazardous substances used at the workplace and put a copy of any MSDS obtained in the register.

— Take reasonable steps to ensure the MSDS is not changed other than by the manufacturer or importer.

— Keep the MSDS close to where the substance is being used.

— Ensure a label is fixed to a hazardous substance container.

— Ensure warnings are given about enclosed systems containing hazardous substances.

Retailers are not required to distribute MSDSs. However, if a hazardous substance is purchased from a retailer, and the substance is for use at a workplace, an MSDS can be requested from another supplier of the hazardous substance such as the manufacturer or importer.

In certain circumstances a supplier must provide copies of the MSDS to the workplace and fix a label to the containers of all classified hazardous substances because the substances:

— are on the National Occupational Health and Safety Commission (NOHSC) List of Designated Hazardous Substances

— on the designated list and are contained in a substance above a certain concentration

— meet the Approved Criteria (PDF, 405 KB) (because of health effects).

More information about MSDS is provided in Section 1 of the Hazardous Substances Advisory Standard 2003.

The format and content for a MSDS used in Australia is set out in the 'National Code of Practice for the Labelling of Workplace Substances'.

Employers can also ask the supplier of a hazardous substance for a 'National Industrial Chemicals

Notification and Assessment Scheme (NICNAS) summary report' which provides more detailed advice about health hazards and control measures.

7.2.1.2. Labelling and decanting

Suppliers, employers and self-employed persons have specific labelling obligations for all hazardous substances containers in the workplace.

The label must be in English and contain the following:

— name of the product

— risk and safety phrases - as stated in NOHSC's document entitled'·

— chemical names of particularly hazardous ingredients

— chemical or generic names of certain other ingredients.

If the manufacturer has amended a MSDS, the label should be changed to ensure that it is consistent with the information in the amended MSDS.

Containers of decanted hazardous substances at the workplace must be labelled with the product name and basic health and safety information (risk and safety phrases) from the supplier's label.

Health effects from hazardous substances in workplaces

It is important when using hazardous materials in the workplace they are properly controlled if they are:

— toxic

— harmful

— corrosive

— irritant

— sensitising

— carcinogenic (causing cancer)
— mutagenic (causing genetic damage)
— teratogenic (causing abnormalities of the foetus).

Some of the health effects of exposure to hazardous materials include:

— skin irritation
— occupational asthma
— systemic chemical poisoning
— chemical burns from corrosives
— cancer.

Factors that determine whether illness or disease occurs include:

— amount and route of exposure
— simultaneous exposure to other hazardous substances
— sensitivity to the substance's effects.

Some of the ways hazardous materials can enter the body include:

— breathing in (inhalation)
— skin contact (where skin is the target organ)
— absorbed through the skin and mucous membranes of the eye
— accidentally swallowed by eating or smoking with contaminated hands
— accidental injection through the skin.

These health effects can be acute, resulting from short-term (usually high) exposure, or chronic, resulting from long term (often low level) exposure over a period of time. Chronic effects may not occur for many years and the cause is often hard to identify.

7.2.1.3. Assessing the risk of exposure

Employers and self-employed persons must manage the risk from worker exposure to hazardous substances in the workplace to prevent serious illness or disease.

The following issues need to be considered to manage the risk when people are exposed to hazardous materials in the workplace:

— assessing the risk

— controlling the exposure

— reviewing health and exposure

— recording risks

— training workers.

Workers must:

— use personal protective equipment (PPE) where it has been supplied by the employer and in the correct manner

— follow instructions given to ensure health and safety

— not wilfully misuse anything provided by the employer to ensure health and safety.

If the workplace has a Workplace Health and Safety Committee, the committee can assist the employer to plan, implement and monitor measures to reduce the exposure to hazardous materials in the workplace. When a hazardous material is to be introduced into the workplace or any changes to the way work is done with a hazardous material, the employer must consult with the committee representative (WHSR).

If an employee is concerned about the way a hazardous material is used in the workplace, they can raise the matter with the committee or the WHSR.

Conducting a risk assessment and keeping a record for hazardous substances

To carry out a risk assessment associated with the use of hazardous materials, employers and self-employed persons need to consider the following:

— the identity of hazardous materials used at the workplace
— a review of the hazardous substance's health effects from the MSDS and label
— how hazardous materials are used
— how people are exposed to hazardous materials
— how exposure to hazardous materials should be controlled
— whether the risk from the hazardous material is significant
— are monitoring or health surveillance required
— a written record of the risk assessment.

Employers can use a generic assessment prepared by an industry body rather than develop one. If an employer does use a generic risk assessment, it must be from a similar type of workplace e.g. a service station should use a generic risk assessment developed for service stations.

7.3. Pandemics and Epidemics

A pandemic is a global disease outbreak. An influenza pandemic occurs when a new influenza virus emerges for which there is little or no immunity in the human population and the virus begins to cause serious illness and then spreads easily person-to-person worldwide.

According to the World Health Organization (WHO), a pandemic can start when three conditions have been met:

— the emergence of a disease new to the population.
— the agent infects humans, causing serious illness.

— the agent spreads easily and sustainably among humans.

A disease or condition is not a pandemic merely because it is widespread or kills many people; it must also be infectious. For example cancer is responsible for many deaths but is not considered a pandemic because the disease is not infectious or contagious (although certain causes of some types of cancer might be).

7.3.1. Pandemic Influenza Phases

The *WHO global influenza preparedness plan* defines the stages of pandemic influenza, outlines the role of WHO and makes recommendations for national measures before and during a pandemic. The phases are:

7.3.1.1. Interpandemic period:

— Phase 1: No new influenza virus subtypes have been detected in humans.

— Phase 2: No new influenza virus subtypes have been detected in humans, but an animal variant threatens human disease.

7.3.1.2. Pandemic alert period:

— Phase 3: Human infection(s) with a new subtype but no human-to-human spread.

— Phase 4: Small cluster(s) with limited localized human-to-human transmission

— Phase 5: Larger cluster(s) but human-to-human spread still localised.

7.3.1.3. Pandemic period:

— Phase 6: Pandemic: increased and sustained transmission in general population.

7.3.2. Major Pandemics

7.3.2.1. Ebola virus and other quickly lethal diseases

Lassa fever, Rift Valley fever, Marburg virus, Ebola virus and Bolivian hemorrhagic fever are highly contagious and deadly diseases with the theoretical potential to become pandemics. Their ability to spread efficiently enough to cause a pandemic is limited, however, as transmission of these viruses requires close contact with the infected vector. Furthermore, the short time between a vector becoming infectious and the onset of symptoms allows medical professionals to quickly quarantine vectors and prevent them from carrying the pathogen elsewhere. Genetic mutations could occur which could elevate their potential for causing widespread harm, thus close observation by contagious disease specialists is merited.

7.3.2.2. Antibiotic resistance

Antibiotic-resistant "superbugs" may also revive diseases previously regarded as "conquered." Cases of tuberculosis resistant to all traditionally effective treatments have emerged to the great concern of health professionals. Such common bacteria as Staphylococcus aureus, Serratia marcescens and species of Enterococcus that have developed resistance to the strongest available antibiotics such as vancomycin emerged in the past 20 years as an important cause of hospital-acquired nosocomial infections, and are now colonising and causing disease in the general population.

In the U.S., 2,000,000 people per year are catching hospital-acquired infections after having been admitted to hospitals to receive medical care for unrelated reasons. The latest number of infections are startling, equating to 4 new cases per minute. Of those, 90,000+ people die.

Organisations like the Center for Disease Control, WHO and Safe Care Campaign are leading the effort to eradicate these avoidable, yet deadly infections.

7.3.2.3. HIV infection

HIV — the virus that causes AIDS — is now considered a global pandemic with infection rates as high as 25% in southern and eastern Africa. Effective education about safer sexual practices and bloodborne infection precautions training have helped to slow down infection rates in several African countries sponsoring national education programs. Infection rates are rising again in Asia and the Americas.

7.3.2.4. SARS

In 2003, there were concerns that SARS, a new, highly contagious form of atypical pneumonia caused by a coronavirus dubbed SARS-CoV, might become pandemic. Rapid action by national and international health authorities such as the World Health Organisation helped slow transmission and eventually broke the chain of transmission, ending the localised epidemics before they could become a pandemic. The disease has not been eradicated, however, and could re-emerge unexpectedly, warranting monitoring and case reporting of suspicious cases of atypical pneumonia.

7.3.2.5. Avian flu

Wild aquatic birds are the natural hosts for a range of influenza A viruses. Occasionally viruses are transmitted from these species to other species and may then cause outbreaks in domestic poultry or (rarely) give rise to a human pandemic.

In February 2004, avian influenza virus was detected in birds in Vietnam, increasing fears of the emergence of new variant strains. It is feared that if the avian influenza virus combines with a human influenza virus (in a bird or a human), the new subtype created could be both highly contagious and highly lethal in humans. Such a subtype could cause a global influenza pandemic, similar to the Spanish Flu, or the lower mortality pandemics such as the Asian Flu and the Hong Kong Flu.

From October 2004 to February 2005, some 3,700 test kits of the 1957 Asian Flu virus were accidentally spread around the world from a lab in the US.

In May 2005, scientists urgently call nations to prepare for a global influenza pandemic that could strike as much as 20% of the world's population.

In October 2005, cases of the avian flu (the deadly strain H5N1) were identified in Turkey. EU Health Commissioner Markos Kyprianou said: "We have received now confirmation that the virus found in Turkey is an avian flu H5N1 virus. There is a direct relationship with viruses found in Russia, Mongolia and China." Cases of bird flu were also identified shortly thereafter in Romania, and then Greece. Possible cases of the virus have also been found in Croatia, Bulgaria and in the United Kingdom. However, by the end of October only 67 people had died as a result of H5N1 which was atypical of previous influenza pandemics.

Despite sensational media reporting, avian flu cannot yet be categorised as a "pandemic" because the virus cannot yet cause sustained and efficient human-to-human transmission. Cases so far are recognised to have been transmitted from bird to human, but as of December 2006 there have been very few (if any) cases of proven human-to-human transmission. Regular influenza viruses establish infection by attaching to receptors in the throat and lungs, but the avian influenza virus can only attach to receptors located deep in the lungs of humans, requiring close, prolonged contact with infected patients and thus limiting person-to-person transmission. The current WHO phase of pandemic alert is level 3, described as "no or very limited human-to-human transmission."

7.3.3. Epidemics

An epidemic is a classification of a disease that appears as new cases in a given human population, during a given period, at a rate that substantially exceeds what is "expected," based on recent experience (the number of new cases in the population during a specified period of time is called the "incidence rate"). (An epizootic is the same thing but for an animal population.)

Defining an epidemic can be subjective, depending in part on what is "expected". An epidemic may be restricted to one locale (an outbreak), more general (an "epidemic") or even global (pandemic). Because it is based on what is "expected" or thought normal, a few cases of a very rare disease like rabies may be classified as an "epidemic," while many cases of a common disease (like the common cold) would not.

Common diseases that occur at a constant but relatively high rate in the population are said to be "endemic." An example of an endemic disease is malaria in some parts of Africa (for example, Liberia) in which a large portion of the population is expected to get malaria at some point in their lifetimes.

Famous examples of epidemics include the bubonic plague epidemic of Medieval Europe known as *the* Black Death, and the Great Influenza Pandemic concurring with the end of World War I. In August 2007, the World Health Organisation reported an unprecedented rate of propagation of infectious diseases.

7.3.3.1. Radiation emergencies

Radiation is a form of energy that is present all around us. Different types of radiation exist, some of which have more energy than others. Amounts of radiation released into the environment are measured in units called curies. However, the dose of radiation that a person receives is measured in units called rem.

7.3.3.2. Terrorism

Terrorism in the modern sense is violence or other harmful acts committed (or threatened) against civilians for political or other ideological goals. Most definitions of terrorism include only those acts which are intended to create fear or "terror", are perpetrated for an ideological goal (as opposed to a lone attack), and deliberately target or disregard the safety of non-combatants. Many definitions also include only acts of unlawful violence.

As a form of unconventional warfare, terrorism is sometimes used when attempting to force political change by convincing a government or population to agree to demands to avoid future harm or fear of harm, destabilising an existing government, motivating a disgruntled population to join an uprising, escalating a conflict in the hopes of disrupting the status quo, expressing a grievance, or drawing attention to a cause.

Terrorism has been used by a broad array of political organisations in furthering their objectives; both right-wing and left-wing political parties, nationalistic, and religious groups, revolutionaries and ruling governments. The presence of non-state actors in widespread armed conflict has created controversy regarding the application of the laws of war.

An International Round Table on Constructing Peace, Deconstructing Terror hosted by Strategic Foresight

Group recommended that a distinction should be made between terrorism and acts of terror. While acts of terror are criminal acts as per the United Nations Security Council Resolution 1373 and domestic jurisprudence of almost all countries in the world, terrorism refers to a phenomenon including the actual acts, the perpetrators of acts of terror themselves and their motives.

There is disagreement on definitions of terrorism. However, there is an intellectual consensus globally, that acts of terror should not be accepted under any circumstances. This is reflected in all important conventions including the United Nations counter terrorism strategy, the decisions of the Madrid Conference on terrorism, the Strategic Foresight Group and ALDE Round Tables at the European Parliament.

Official definitions determine counter-terrorism policy and are often developed to serve it. Most government definitions outline the following key criteria: target, objective, motive, perpetrator, and legitimacy or legality of the act. Terrorism is also often recognisable by a following statement from the perpetrators.

Violence – According to Walter Laqueur of the Center for Strategic and International Studies, "the only general characteristic of terrorism generally agreed upon is that terrorism involves violence and the threat of violence." However, the criterion of violence alone does not produce a useful definition, as it includes many acts not usually considered terrorism: war, riot, organised crime, or even a simple assault. Property destruction that does not endanger life is not usually considered a violent crime, but some have described property destruction by the Earth Liberation Front and Animal Liberation Front as violence and terrorism.

Psychological impact and fear – The attack was carried out in such a way as to maximise the severity and length of the psychological impact. Each act of terrorism is a "performance," devised to have an impact on many large audiences. Terrorists also attack national symbols to show their power and to shake the foundation of the country or society they are opposed to. This may negatively affect a government's legitimacy, while increasing the legitimacy of the given terrorist organisation and/or ideology behind a terrorist act.

Perpetrated for a Political Goal – Something all terrorist attacks have in common is their perpetration for a political purpose. Terrorism is a political tactic, not unlike letter writing or protesting, that is used by activists when they believe no other means will effect the kind of change they desire.

The change is desired so badly that failure is seen as a worse outcome than the deaths of civilians. This is often where the interrelationship between terrorism and religion occurs. When a political struggle is integrated into the framework of a religious or "cosmic" struggle, such as over the control of an ancestral homeland or holy site such as Israel and Jerusalem, failing in the political goal (nationalism) becomes equated with spiritual failure, which, for the highly committed, is worse than their own death or the deaths of innocent civilians.

Deliberate targeting of non-combatants – It is commonly held that the distinctive nature of terrorism lies in its intentional and specific selection of civilians as direct targets. Much of the time, the victims of terrorism are targeted not because they are threats, but because they are specific "symbols, tools, animals or corrupt beings" that tie into a specific view of the world that the terrorist possess. Their suffering accomplishes the terrorists' goals of instilling fear, getting a message out to an audience, or otherwise accomplishing their political end.

Disguise – Terrorists almost invariably pretend to be non-combatants, hide among non-combatants, fight from in the midst of non-combatants, and when they can, strive to mislead and provoke the government soldiers into attacking the wrong people, that the government may be blamed for it. When an enemy is identifiable as a combatant, the word terrorism is rarely used. Mass executions of hostages, as by the Nazi military forces in the Second World War, certainly constituted crimes against humanity but are not commonly called terrorism.

Unlawfulness or illegitimacy – Some official (notably government) definitions of terrorism add a criterion of illegitimacy or unlawfulness to distinguish between actions authorised by a "legitimate" government (and thus "lawful") and those of other actors, including individuals and small groups. Using this criterion, actions that would otherwise qualify as terrorism would not be considered terrorism if they were government sanctioned. For example, firebombing a city, which is designed to affect civilian support for a cause, would not be considered terrorism if it were authorised by a "legitimate" government.

This criterion is inherently problematic and is not universally accepted, because: it denies the existence of state terrorism; the same act may or may not be classed as terrorism depending on whether its sponsorship is traced to a "legitimate" government; "legitimacy" and "lawfulness" are subjective, depending on the perspective of one government or another; and it diverges from the historically accepted meaning and origin of the term. For these reasons this criterion is not universally accepted. Most dictionary definitions of the term do not include this criterion.

Types of terrorism

Terrorism classified terrorism into six categories.

— *Civil Disorders* – A form of collective violence interfering with the peace, security, and normal functioning of the community.

— *Political Terrorism* – Violent criminal behaviour designed primarily to generate fear in the community, or substantial segment of it, for political purposes.

— *Non-Political Terrorism* – Terrorism that is not aimed at political purposes but which exhibits "conscious design to create and maintain high degree of fear for coercive purposes, but the end is individual or collective gain rather than the achievement of a political objective."

— *Quasi-Terrorism* – The activities incidental to the commission of crimes of violence that are similar in form and method to genuine terrorism but which nevertheless lack its essential ingredient. It is not the main purpose of the quasi-terrorists to induce terror in the immediate victim as in the case of genuine terrorism, but the quasi-terrorist uses the modalities and techniques of the genuine terrorist and produces similar consequences and reaction. For example, the fleeing felon who takes hostages is a quasi-terrorist, whose methods are similar to those of the genuine terrorist but whose purposes are quite different.

— *Limited Political Terrorism* – Genuine political terrorism is characterised by a revolutionary approach; limited political terrorism refers to "acts of terrorism which are committed for ideological or political motives but which are not part of a concerted campaign to capture control of the State.

— *Official or State Terrorism* -- referring to nations whose rule is based upon fear and oppression that reach similar to terrorism or such proportions."

Causes

Many opinions exist concerning the causes of terrorism. They range from demographic to socioeconomic to political factors. Demographic factors may include congestion and high growth rates. Socioeconomic factors may include poverty, unemployment, and land tenure problems. Political factors may include disenfranchisement, ethnic conflict, religious conflict, territorial conflict, access to resources, or even revenge.

Factors that to terrorism

—High population growth rates (so-called "youth bulges")

— High Unemployment

— Lagging economies

— Political disenfranchisement

  — Extremism

  — Ethnic conflict

  — Religious conflict

  — Territorial conflict

In some cases, the rationale for a terrorist attack may be uncertain (as in the many attacks for which no group or individual claims responsibility) or unrelated to any large-scale social conflict (such as the Sarin gas attack on the Tokyo subway by Aum Shinrikyo).

A global research report *An Inclusive World* prepared by an international team of researchers from all continents has analysed causes of present day terrorism. It has reached the conclusions that terrorism all over the world functions like an economic market. There is demand for terrorists placed by greed or grievances. Supply is driven by relative deprivation resulting in triple deficits - developmental deficit, democratic deficit and dignity deficit. Acts of terror take place at the point of intersection between supply and demand. Those placing the demand use religion and other denominators as vehicles to establish links with those on the supply side. This pattern can be observed in all situations ranging from Colombia to Colombo and the Philippines to the Palestine.

8

Vulnerability Analysis

Historically there has been a split in disasters work which sees these two - hazards and vulnerability - as separate arenas, each with their own specialists. Inherent in this is the danger that those who specialise in dealing with hazards tend not to deal with vulnerability (let alone the interaction). Many hazard specialists also tend to deal in one type of hazard, and to be rooted in a physical science where knowledge of (or even interest in) the social sciences is minimal.

At the same time, with the emergence of a wider awareness of vulnerability issues, there is also the danger that it is being used crudely or simplistically, and incorporated into disaster work in depoliticised and inadequate ways. This is particularly significant as the word itself implies people being potential victims, in need of assistance and incapacitated. For instance, the British Red Cross use the word to mean 'people in need and crisis', which is little different from the term 'victim' and has no sense of prediction. It is therefore crucial to recognise that vulnerability is balanced by peoples' capabilities and resilience, and that if they are perceived only or mainly as victims then the problem of what causes vulnerability may be evaded.

Another problem is that vulnerability analysis risks being regarded as politically neutral and devoid of contention and conflict, its connotation being that it is simply about incorporating people into the equation in a more prominent manner. If this is the case, then vulnerability has now become one of those buzzwords akin to 'sustainability', used in so many contexts that it is in danger of becoming useless.

This makes an attempt to specify and operationalise what is meant by vulnerability particularly important. We need to disaggregate it, make it apparent that it is derived largely from a political, economic and social context and is not simply about people who are 'victims' in some aggregated and apolitical manner. The key issue is that people's own (very variable) characteristics - their capacities, resilience and vulnerabilities - are recognised as a significant part of the disaster equation. Moreover, vulnerability needs to be appraised in terms of the differential impacts of various types of hazard impacts, and operationalised so that the factors that constitute vulnerabilty can be measured and taken into account prior to a hazard striking.

Vulnerability analysis is developed from a range of socio-economic approaches to hazards and what we could call 'the disaster of everyday life'. Pelling's work on floods in Guyana sees vulnerability as 'an ongoing state rather than a status to be identified in relation to a specific hazardous event'.

In other words, vulnerability analysis begins with the crucial acceptance that vulnerability is often part of the normal, becoming apparent and obvious to some only with the impact of a hazard. It overlaps with and is derived from other perspecitves including (among others) Amartya Sen's work on famine and entitlements, the UK Save the Children Fund's project on famine warning through the software system RiskMap and a great deal of other work on food systems and coping strategies.

It is vital to recognise that vulnerability should be treated as a condition of people that derives from their political-economic position. It is therefore 'dangerous' to use it loosely or as a characteristic of exposure to hazards alone, since this allows for the key components of power and income distribution to be played down and prominence given to technical fixes.

The assumption that 'natural' disasters are inherently and predominantly natural phenomena has tended to exclude the social sciences from consideration in much of the spending that is done in disaster preparedness. This is despite the fact that over the last twenty years a considerable literature on disasters has emerged from human geography, sociology, anthropology and (to a lesser extent) economics. For many years, social science has contributed to policy formation for disasters (especially in the Third World) through the activities of many Non-Government Organisations (NGOs).

The initial development of vulnerability analysis is then rooted in social science, and in a sense has constituted a political economy of disasters to the analysis of devastating events that are normally associated with natural hazards. At its most simplistic, vulnerability analysis asserts that for there to be a disaster there has to be not only a natural hazard, but also a vulnerable population. Much of the conventional work on disasters has been dominated by 'hard science', and has been a product of the prominence that natural phenomena have acquired in the disaster causation process. But this 'physicalist' approach is also a result of the *social construction of disasters* as events that demonstrate the human condition as subordinate to Nature.

Within such a framework, there is the inherent danger that people are perceived as victims rather than being part of socio-economic systems that allocate risk differently to various types of people. People therefore often become treated as 'clients' in the process of disaster mitigation and preparedness, and as passive onlookers in a process in which science and technology do things to them and for them, rather than with them.

In the last five years or so, the term vulnerability analysis has become more widely used, and in some disaster disciplines the notion of vulnerability has become common. The notion is that the analysis of the vulnerability (of people and not only physical structures) would allow some measure of mitigation and preparation, if not socio-economic restructuring. So a key element of the term is that it should be prescriptive and predictive. The focus should be on its political economy determinants and their effects in differentiating people (into groups that are differentially exposed to risk), and not simply structures that happen to be in places where a particular hazard (or various hazards) is likely to strike.

Most usages of the idea of vulnerability accept that it is part of a continuum or ranking of people, and that being vulnerable is at the 'negative' end of such a scale. Granger has suggested that 'vulnerability of each element at risk within the community can be measured along a contunuum from total resilience at one end to total susceptability at the other.' The term in its political economy usage implies that while there can be a ranking of people from more to less vulnerable (with capabilities and resilience at the positive end), the continuum must be related to various political, social and economic components of vulnerability, and that there can be different types of vulnerability according to the different hazards that might affect a given place (for instance a particular family may be more vulnerable to wildfire than to earthquake in the same place.)

This means that vulnerability analysis is complex and dependent on large data sets, and on qualitative analysis that requires the involvement of the people concerned in the evaluation of their vulnerability. The focus is either on groups of people who prima facie are vulnerable in the sense that they are clearly low on all or most socio-economic indicators (a 'disaster waiting to happen'), or in places where conventional civil defence approaches are seen as inadequate, and community-led responses are possible. In the first type of situation, examples include the work of Intermediate Technology in Peru, and the Central America survey of the early 1990s. Much of the most innovative work is going on in Third World countries, where NGOs have become aware of the restrictions of the 'hard science' approach.

Vulnerability can be considered on a scale from high to low levels for a number of components. These components recognise not only the negative ends of the scale (vulnerabilities), but also how they can constitute the positive capabilities of an individual or group to survive and recover from a given hazard impact of a given severity.

Vulnerability can be considered in terms of five components: Initial well-being, Self-protection, Social protection, Livelihood Resilience, and Social Capital. It should be noted that each one of these is crucially linked to the likely severity of impact of a given hazard, and yet primarily they are all determined by political, economic or social processes. Each of these contains the possibility of both vulnerabilities and capabilities, with these varying over time (as individuals and groups subsist and compete within given livelihood possibilities), and being affected in regard to different types of natural hazards. Of course, these components are also part and parcel of everyday life, and are not only related to the likely (or unlikely) impacts of different natural hazards. They are also of relevance to a person or group's ability to withstand (or be involved with) other forms of short-term shock or unforseen circumstances (such as civil conflicts and war, or Man-made hazards, or complex emergencies which combine either or both of these with natural hazards).

The five components of the level of vulnerability are then:

1. *Initial well-being, strength and resilience*: This evaluates the initial nutritional and health status (both physical and mental) of people in everyday life (or before the impact of a hazard). It is indicative of their capacity to cope with illness and some types of injury resulting from a hazard. It should include their potential for mental disturbance and recovery in the wake of a disaster, which might intensify existing stresses. A person's resilience may relate to having a faith or spiritual confidence, or a predisposition to self-reliance.

2. *Livelihood resilience* : A measure of the capacity of an individual and/or their household to cope with the aftermath of a given hazard impact, and to reinstate their earning or livelihood pattern. This might include their likely continued employment, level of savings, loss of welfare benefits, loss or injury of supportive family members, hazard damage to their normal livelihood activity (for example in floods this might include damage to agricultural land by sediment deposits, sea-water incursion, toxic or sewage contamination).

3. *Self-protection concerns the ability* or willingness of an individual and/or household (with a given level of knowledge of apparent risks) to provide themselves with adequate protection, or to be able to avoid living or working in hazardous places. It will be influenced by the level of knowledge of physical measures, and the capacity of people to implement them.

4. Societal protection refers to the ability or willingness of social and political structures at political or social levels above the individual or household, to provide protection (especially structural and technical preparations) from particular hazards. This might include local government, national government, relevant organisations (e.g. fire department, civil defence), or community-based initiatives.
5. Social capital involves the 'soft' security provided by group or community capacities to enhance (or reduce) a person's resilience. This may include the degree of cohesion or rivalry that might affect rescue and recovery. There are various forms of social capital that may enhance or hinder recovery. These include support networks (belonging to a church or other group), some of which may provide mutual aid in times of hardship. The character and quality of social capital may depend to a large extent on the type of state power and the capacity for civil society to develop

As can be seen, these five components place someone in the spectrum from highly vulnerabe to being secure and are a complex mix of an individual's characterstics, social factors and economic and political processes, and the type of hazard to which they might be exposed. They involve both 'hard', generally hazard-specific technical interventions like warning systems and physical structures like cyclone shelters and flood embankments, and 'soft' socio-economic factors (including income distribution, access to livelihood resources, discrimination in the receipt of assistance). Various processes and factors determine the extent to which a person or group is made vulnerable or secure in relation to each of these components of vulnerability.

These components must then be cross-related to a series of social factors and political characteristics that can be considered to affect them positively or negatively (determinants), so generating different levels of vulnerability for each of the components. The social factors include economic class (or income group as its surrogate), gender, ethnicity, and age. The political characteristics of different societies can be interpreted in four ways: firstly the type of state system (democratic, redistributive, pro-corporation, authoritarian, kleptocratic, religious, etc.); secondly the state's capacity to act (its 'reach', whether it has adequate revenue, its efficiency); thirdly the strength of civil society that the state enables or permits; and fourthly other factors which lead to cohesion or cleavages, such as whether a participatory society is fostered or forbidden, how much dependency there is on religious or political allegiances, and how these are organised in opposition or support for the state.

These social factors also affect the way that scientific and technical knowledge of hazards (and how to prepare for them) is used, how good it is and how that science and technology is 'distributed' between different groups of people. In other word, it is common to talk about income and asset distribution and the way these affect different groups of people. But we can also speak about 'scientific and technical distribution' as well, since it can operate within power relations that make knowledge unequally available to different types of people, leaving them more or less vulnerable.

Vulnerability is is also difficult to operationalise, as the data requirements are high, and some of the variables themselves (for instance the type of state power, capabilities in civil society, type of science operated) may make vulnerability analysis difficult to implement if they reduce political will. Since vulnerability analysis is relatively new and little has been done in data collection and analysis, some of the statements are really hypotheses which require testing.

It should also be noted that there is no simple correlation between someone enjoying a low level of vulnerability and the normal 'advantaged' conditions of the rich, of men, or of dominant ethnic groups. There are occasions when being rich or male can generate more exposure to hazard risk, and vulnerability is not necessarily the same as poverty or marginalisation, though it appears that in most cases there is a reasonable parallel.

8.1. Components and Variables

8.1.1. Initial Well-being

This has a strong positive correlation with class (income) level, such that nutritional status and the resulting resistance to disease and level of mental capacity is higher in better-off people. High levels of well-being are likely to reduce the risks from disease vectors after floods, and to permit more rapid recovery from injury. Given that females in the Third World normally have worse nutritional status than males, and in many cultures are less likely to have equal access to health care, gender is an important factor in generating unequal well-being.

Minority ethnic groups who suffer discrimination may have lower nutritional status and may live in poorer accomodation which has made them less healthy. The type of state power may vary from one with good levels of welfare distribution and health care which can enhance the nutritional status of lower classes and ethnic minorities, to others which institutionalise gross income inequalities and prevail over mass undernutrition (e.g. Brazil, India). But even states which are 'democratic' and espouse welfare state ideals can be in command of countries where ethnic minorities (e.g. the USA, Australia) or even ethnic majorities (e.g. South Africa under apartheid; Guatamala's Amerindians) are malnourished or subject to cultural disintegration and poor morale. The type of state often has a close relationship with the prospects for civil society and its potential for a positive contribution in building peoples capacity to deal with crises of various types. In regard to well-being and nutrition, a healthy civil society is likely to foster a more healthy population, as a result of the circulation and sharing of information, an openness to debate about health issues, and the potential safeguard of a free press against famine and other extremes.

8.1.2. Livelihood Resilience

Poorer people in lower classes may have less job security after a flood, and lower levels of savings to buffer them against the shock. In rural employment in the Third World, floods may reduce demand for labour by destroying crops on which people normally work, and in towns and cities the opportunity for earning in the formal and informal sector is likely to be interrupted or reduced. Some people may gain new earning opportunities, for example providing boat transport or vending services to marooned people, as happens in Bangladesh. In general though, it is to be expected that many livelihoods are disrupted and that the impact is disproportionately bad for those who are already poor. They are also unlikely to have insurance, so that there is a double loss in the sense that many household goods (or the house itself) may be lost so that when income does become earned again, it is having to recoup other losses as well as provide basic needs for survival.

There may also be negative gender consequences if women find it more difficult to re-enter work compared with men. Women who are dependent on men in the household may also be kept in more passive 'victim' roles, with less of a voice in the recovery process in its male-gendered management structure. On the opposite side, the gendered professional care system is predominantly feminised, and this may alienate men from seeking help, reducing their ability to recover. There are also indications that men's mental health can be affected badly (to the point of suicide) by their own 'gendered' feelings of failure to remain a 'proper' provider for the family in the aftermath of a disaster, as in the case of flood victims in Australia in 1993. This may also apply to ethnic minorities, who may find that they are lower down the priorities in being given employment.

8.1.3. Self Protection

In relation to class inequalities, there are some anomalies concerning self-protection. While it appears superficially that higher income groups will be able to afford better-built houses and the avoidance of unsafe work-places, it is not so straightforward. In the Kilari – Latur earthquake for example many of the victims of house collapse were from the wealthier families. Heavy stone homes had been constructed which buried victims and caused higher mortalities and injuries than among poorer people. In parts of California, wealthy people lose their homes to landslides and wildfires, but this is largely because the locations are chosen for other advantages (of neighbourhood or vista), and so the risks are both known and insured against.

For gender many factors may play a part in disadvantaging women. Place and type of work may put women (and children who are with them) more at risk. Seemingly simple things like being able to swim may make a difference, and it has been argued that prejudice against women learning to swim in Bangladesh has had a significant effect on their ability to survive in river and cyclonic floods.

Regarding ethnic issues, income-earning capacity and prejudice may reduce the capacity for some groups to provide themselves with safe buildings, and may lead them to be working in more dangerous places or more marginal land which produces less livelihood security. The type of state may affect people differently in terms of their ability to avail themselves of relevant techniques of safe construction or hazard preparedness. For instance, the Pakistan government allowed and encouraged the migration of skilled workers to the Arabian Gulf in the 1970s, in order to gain from their foreign exchange remitances. As a result there was a shortage of carpenters with the skills necessary for safer house construction in the earthquake and landslide regions of the Karakoram.

8.1.4. Societal Protection

Women may also be victims of gender prejudice when societal protection is targeted at the household, as males often have preferential access to relief supplies and the opportunities given by rehabilitation programmes. Fordham argues that disaster management itself is male gendered, embodying a weak understanding of women's lives and an assumption of male-headed households (with men commanding access to relief and other welfare) and no differentiated female capacities or vulnerabilities.

Ethnic groups which suffer discrimination may not be granted the same access to warning systems or other precautionary measures. In California, Mexican migrant workers are mostly excluded from medical care and welfare after earthquakes and other hazards, and even when they have residence status are often afraid to seek help because of prejudice or the danger of being repatriated. In floods in Victoria, Australia there was evidence that non-English speaking miniority groups suffered as they had difficulty accessing services (but that their communities and families were more supportive - a compensation by social capital).

The state system is highly significant for the type and effectiveness of the precautionary measures that it instigates (or fails to initiate). This affects not only the quality of emergency preparedness, but also the implementation of engineering standards, the effectiveness of building controls and regulations about land zoning, the avoidance of flood-prone areas for settlements, and the trust that the state enjoys in prompting mass evacuations and other procedures. The state also acts as the insurer of last resort in the sense that for uninsured people their only potential 'insurance claim' for post-disaster assistance is to the state and its agencies. The state may even restrict victims' access to other forms of relief provided for instance by charities (foreign and domestic), as was often the case with communist governments.

8.1.5. Social Capital

Social capital can include people's 'cultural resources' and educational level, and things like their ability to deal with bureaucracies. A study of two communities affected by the Northridge earthquake in California found that each fared very differently in terms of access to relief and the success of recontruction. 'Social capital' was higher in the town that did better, and ethnic and class issues were involved in reducing the success of the other. In terms of class, it is likely that poorer groups will be disadvantaged. Social capital also involves people's ability to have access to (or contribute to) various networks and systems of mutual support in times of crisis are an important factor in dealing with hazards. Most assistance in disasters has been shown to be supplied by the affected communities themselves.

Is this capacity improved by higher levels of social cohesion and low levels of inter-group rivalry? Sometimes it appears that some networks that constitute social capital are substitutes for what the state provides inadequately or not at all. After the 1992 earthquake that affected Cairo, there is evidence that the Muslim Brotherhood was able to operate more effectively than the government in organising rescue and relief, and that it was able to gain political support as a result.

Sometimes ethnic prejudice may result in some people not taking notice of the help available from others. There was some evidence when a tropical cyclone hit the Darwin area (north Australia) in1974 that the local Aborigines had left in the knowledge that a dangerous storm was imminent, while the other people remained until evacuation was difficult and inadequate.

Gender components of social capital may be relevant, as it is possible that women play a more significant role in creating and nurturing the linkages. The state may influence the emergence of social capital by restricting people's ability to organise, or by fostering inter-group rivalries that mean that assistance is denied to some groups.

8.2. Vulnerability Modelling

The dominance of hard science in work on disasters has tended to mean that most mitigation proposals are dominated by a 'technical fix' approach. These tend to address only limited components of peoples' vulnerability, mainly in Societal Protection and in providing the technical capability (but often not the means for implementation) for Self-protection. In other words, and especially in Third World contexts, dealing with peoples' livelihood resilience is not regarded as susceptible to technical interventions and so is 'defined out' of the problem. Moreover, since the main components of what causes these forms of vulnerability is governed by politics and economics, vulnerability analysis is avoided as being 'not relevant to science' or 'too difficult to get involved in'. In effect, what happens then is that vulnerability is addressed only in aspects that are susceptible to technical interventions, but because the main causes are ignored these interventions themselves sometimes reinforce the conditions that generate vulnerability.

In relation to riverine and rainfall floods, technical interventions have usually meant storm drains and channel modifications in urban environments, and river training and embanking elsewhere. Although the issues of land modification and changes to runoff from built-up areas have led to policies of land-use zoning to avoid e.g. flood plain risks, it seems clear that commercial pressures or inadequate implementation in some countries has reduced the efficacy of such engineering. And the issue of river training and embanking has become extremely controversial, especially after floods of the Mississippi, and the Rhine and its tributaries in Germany and the Netherlands in recent years.

But the issue is not really about whether Nature can be 'controlled' and subdued or not, but the type of control and set of choices that are presumed to be available within a given socio-economic system. In most capitalist and communist contexts, this has generally meant an overwhelming focus on practices that emerge from and reinforce that socio-economic system, rather than being able to think about how real people with actual vulnerabilities are interacting with hazards. As a result, huge capital investments are made in river training schemes with little consideration for the opportunity costs and how that capital might be spent in other ways to deal with the forces that generate peoples' vulnerability.

The environmentalist arguments about restoring rivers to flow more naturally and accept that people should 'live with floods' may well redress the errors of ineffective capital spending on hardware approaches. But they do not necessarily deal with the political economy of people and their vulnerabilities. The same must be said of vulnerability analysis: while it may be technically possible to do it, how are the causes of peoples' vulnerabilites - the political, economic and social roots of it - going to be addressed? It is largely because power structures want to avoid dealing with such issues that the 'tech-fix' approach is so dominant.

Modelling the vulnerability of people will require the collection and analysis of data in quantitative and qualitative terms, with a combination of surveys of households, institutional analysis (local governments, insurance companies, voluntary organisations, businesses and employers), livelihood and welfare analysis (of income sources and employment patterns), and surveys of physical structures and infrastructure (with the emphasis not only on property damage, but also the impact of floods on welfare and income earning opportunities). Attempts to design methodologies for this are being made, especially in Australian disaster management. Cross-cultural analysis of vulnerability and potential losses are also being worked out.

The vulnerability modelling will include not only the area at threat of inundation with given flood scenarios, but also surrounding areas that may suffer various forms of disruption for other categories of vulnerable people. For instance, loss of a significant employer through building damage will cause not only a loss to the business (and the insurer), but also widespread disruption of livelihoods and earning capacity for employees, whose vulnerability may be high. On the other hand, some communities may have more 'social capital' in the form of local organisations that enable people to recover more quickly than elsewhere. Although such organisations are not designed to deal with floods, they may permit greater social cohesion and higher morale.

8.3. Dimensions of Vulnerability

Social, generational, geographic, economic and political processes that influence how hazards affect people in varying ways and with different intensities. Some groups are more prone to damage, loss and suffering in the context of differing hazards. Key variables explaining variations of impact include class, occupation, caste, ethnicity, gender, disability and health status, age and immigration status and the nature and extent of social networks. Changing the social, economic and political factors usually means altering the way that power operates in society.

The relative contribution of geophysical and biological processes on the one hand and social, economic and political processes on the other to vulnerability varies from disaster to disaster, as well as from one community to another and from one place to another. The highlighted paragraph illustrates some of the dimensions of vulnerability. Vulnerability can be increased through entitlements, political powerlessness or social exploitation and discrimination. The interactions of the different factors of vulnerability will determine people's capacities, access to resources and ability to realise their rights.

In the face of a particular hazard, it is important to determine how each hazard interacts with each and every dimension of vulnerability. Response planning should be based on the manifestation of vulnerability affecting communities in a particular context.

8.3.1. People and Vulnerability

Vulnerability is gender differentiated. The way women experience vulnerability is many times different to men due to socially constructed gender roles and power relations. Factors, such as lack of access to and control over basic resources and lack of entitlements, amplify women's vulnerability and undermine their ability to cope with effects of disasters.

Some groups of people tend to be more vulnerable than others: children, the elderly and people with disabilities are more vulnerable because of physical difficulties. For other groups, their identity (dalits, ethnic minorities) leads them to exclusion, including people with HIV/AIDS, for a combination of these reasons (physical, financial, stigma, generation). Therefore a study of vulnerability is a study of what might happen to people or communities. While it is not certain that a crisis will happen, it is certain that some people are more likely to be severely affected if a crisis does happen.

8.3.2. Vulnerability and Poverty

Poverty is not the same as vulnerability, but they are strongly linked. They are mutually re-enforcing and brought about by similar processes. All poor people are vulnerable but not all vulnerable people are poor. Poverty is a core dimension of vulnerability. Poverty is not the only factor that leads to vulnerability; other factors like geographical location, communal conflict or social and ethnic association can make people vulnerable. Vulnerability pushes people into poverty, keeps them in poverty and stops them from coming out of poverty.

Poverty is the state of deprivation (lack of access) to key resources necessary for full participation in economic and social life. It is thus thought about as current status, and often heavily associated with material/social status. Increasingly poverty is seen as a multidimensional issue, involving lack of access to a wide range of natural social, economic, and political resources and capacities. Vulnerability, on the other hand, is more about defencelessness, insecurity, exposure to hazards or shocks and the ability to cope with them than it is about current status.

8.3.3. Vulnerability and Human Security

Enquiring about the conditions (e.g. a hazardous physical environment, or severe economic deprivation) that make some people more vulnerable than others is central to PVA. But we need to probe further. We need to probe further. Understanding vulnerability requires a closer scrutiny of the power relations that determine, for instance, who in any given society gets what, who makes decisions and who is excluded. In this respect, PVA draws heavily on our evolving rights- based analysis.

The notion of human security provides us with a useful framework to analyse the links between vulnerability, power and rights. Let us concentrate on the following examples.

About 20,000 people (perhaps several thousand more) died in Orissa in 1999. Hit by a super-cyclone, entire communities were wiped out, two million houses were reduced to rubble, thousands of heads of livestock were killed, and 1.3 million hectares of paddy crops were destroyed. The disaster took the majority of the inhabitants of area by surprise. Yet the fact is that meteorologists had warned that a super-cyclone was brewing in the Bay of Bengal four days before it hit Orissa. Informed about the impending disaster, the Chief Minister of the State took action and consulted three astrologers, who assured him that the cyclone would weaken or be deflected. Unfortunately, the cyclone struck the coastal areas of the state. By the way, cyclones hit Orissa every year.

The 2002 food crisis in Malawi resulted in several hundred hunger-related deaths – perhaps several thousand. Following two good production years, localised floods reduced the maize harvest. Surprisingly the government had already sold much of the strategic grain reserve. As a result, maize prices escalated to unaffordable levels, meaning that when people's crops failed they could not obtain maize, the major staple. This increased their susceptibility to risk. The deepening crisis mostly affected poor rural dwellers, and amongst these people the elderly, young children and the sick.

Throughout the nine-year Sierra Leonean conflict there was wide spread and systematic sexual violence against women and girls including individual and gang rape and sexual slavery. In thousands of cases, sexual violence was followed by the abduction of women and girls and forced bondage to male combatants, often accompanied by forced labour. While members of the rebel groups were the most common perpetrators, members of the civil defence forces and the loyal Sierra Leonean Army were also implicated.

Of course, the reality is far more complex than this. Poor and marginalised people are often exposed to events largely beyond their control (natural hazards, economic crises, violent conflicts, chronic destitution, HIV/AIDS). Incapable of protecting themselves, their security becomes dependent on the protection strategies and mechanisms set up by states, international agencies, NGOs and increasingly the private sector.

The implications of this approach may produce innovative policy approaches. But a key problem is that for vulnerability analysis to be implemented seriously requires the state to engage in activities that deal with inherent inequality and prejudice. Where the state is part of the problem in maintaining such power systems, then it is difficult to see how it can foster adequate solutions to vulnerability. Having said that, it is also possible for some states to recognise the benefits of a vulnerability analysis approach, since there are few governments in the world which officially claim that they are uninterested in protecting their own citizens, and which would not be interested in potentially cheaper ways of reducing their vulnerability.

For example, vulnerability analsyis may suggest that new combinations of flood forecasting and social engineering are more appropriate than flood prevention works based on civil engineering. This may include the encouragement of behavioural changes in communities that face possible flooding, the institution of precautionary activities, community insurance policies (that do not depend on individual ability to pay), and other measures (that foster civil society to enhance social capital) which may be more appropriate and effective than expensive 'hard' flood defences. In this, the focus can be on vulnerable people rather than the hazard itself.

In particular, communities of vulnerable people can themselves be seen as being more active agents of preparedness and mitigation, rather than the passive recipients of policies implemented on their behalf. Valdes demonstrates that flood warning systems based on community operation and participation in Costa Rica 'make a difference whether early warnings are acted upon to save lives and property'.

8.4. Participatory Vulnerability Analysis

PVA helps us identify critical gaps in people's infrastructure of protection. It may be lack of co-ordination, or the doings or neglect of unaccountable and corrupted governments, or the absence of effective enforcement instruments to protect people from violence and aggression, or a combination of all of them. The fact is that human insecurity is often linked to the denial of rights and freedoms and the lack of adequate protection mechanisms (or the failure of existing ones).

PVA also invites us to explore ways to strengthen or improve people's infrastructure of protection. Involved in the process of formulation and implementation of protection strategies, communities may decide to adopt a variety of actions and strategies. For instance, their organisation and mobilisation to put pressure on their government to set up early warning systems; or to claim their right to return to the (safer) land from which they were unlawfully evicted; or link with others at wider levels to urge the international community to stop gross violations of human rights against civilians, particularly women. They may also decide to initiate judicial actions to access their entitlements (e.g. compensation packages). Different contexts will require different strategies and actions.

These examples, however, show that protection is essential, but not enough. A PVA approach promotes empowerment, that is, the active participation of people and communities in determining their wellbeing and building their infrastructure of protection.We know that for poor people to be able to cope, they must have the possibility of developing their individual and collective capabilities to make informed choices and to act on behalf of themselves and others. It may happen through collective action and engagement in public debates and decisionmaking processes (e.g. social audits, legitimate representation in local and national government). It may also result in communities reaffirming their rights and challenging discrimination and exclusion, or taking charge of mediation and peace-building efforts.

By assisting practitioners and communities to enhance people's protection and promote empowerment, PVA offers a valuable contribution to our efforts to reduce vulnerability and attain human security. To sum up, PVA builds on the recognition that everybody has fundamental rights established in different legal and policy instruments as well as cultural codes (e.g. to life and health; to humanitarian assistance; freedom from slavery and sexual violence). It also considers who in particular has what obligations, and who is in a position to help reduce insecurities in human lives.

8.4.1. Analysing Vulnerability

The nature of vulnerability is dynamic and complex and therefore cannot be analysed directly. Hence any assessment or analysis of vulnerability is a predictive judgement – it predicts what's likely to happen and why. The analysis breaks down the detail to the point that it can be understood and addressed. The analogy on the right gives some insights on the complexities of vulnerability.

Various approaches to analysing vulnerability exist: quantitative (measuring vulnerability using quantifiable characteristics, for example 50 people are likely to be affected by landslides in Bundibugyo) and qualitative (analysing vulnerability using characteristics, for example 50 people who are likely to be affected by landslides in Bundibugyo are households who lost their land during the war and who were resettled on the river banks).

PVA is a systematic process that involves communities and other stakeholders in an in-depth examination of their vulnerability, and at the same time empowers or motivates them to take appropriate actions. The overall aim of PVA is to link disaster preparedness and response to long-term development.

PVA is a qualitative way of analysing vulnerability, which involves participation of vulnerable people themselves. The analysis helps us to understand vulnerabiiity, its root causes and most vulnerable groups, and agree on actions by, with and to people to reduce their vulnerability. By analysis we mean the process of breaking down something into component parts, which can then be addressed. PVA has its own principles, which are outlined below.

8.4.2. Principles of PVA

— Active agency – that poor people can and must be

involved in finding the solutions to the problems they face.

— PVA is not an end in itself, it should result in action and change for the better.

— The sources of vulnerability and solutions to vulnerability are located or controlled outside the community, so you need a multi-level process.

— It is based on ActionAid's rights based principles.

PVA uses a step-by-step approach to systematically analyse the causes of vulnerability by:

1. Tracking hazards to determine the level of exposure to risk, causes and effects.
2. Examining unsafe conditions (factors that make people susceptible to risk at a specific point in time).
3. Tracking systems and factors (dynamic pressures) that determine vulnerability, resilience and root causes.
4. Analysing capacities and their impact on reducing vulnerability.

The factors and conditions that cause vulnerability are always changing and progressing, if they are not stopped. From the example on the next page, PVA tracks how vulnerability is progressing over time using the step-by-step approach. On the basis of these changes, provisions can be made when developing policies or programmes to protect people and build their resilience.

8.4.2.1. Multi-levelled approach

There are multiple determinants/causes of vulnerability. Some of these fall outside individuals or community. It follows that analysis of vulnerability should go beyond the individual to micro and macro level political processes. This highlights the need for a multi-levelled approach. Much as community perceptions are vital for developing policies that reduce vulnerability, existing policies will change and protect the most vulnerable if policy makers hear the analysis of those who are vulnerable. The levels are:

1) *Community level:* PVA enables communities to play a dual role, as informants, but also analysts, by breaking down vulnerability to a point where they can begin to take action to reduce their own vulnerability. The analysis itself has no value unless it is followed by action: people can take action themselves or get support. Right solutions to vulnerability cannot be imposed externally but require people who are vulnerable to be involved. Reasons behind vulnerability are seldom as straightforward as they may appear to others. Therefore the community's analysis needs to inform policies and actions. This is a unique way of allowing poor and marginalised people to have a say in policies that affect them.

2) District level: to analyse causes of vulnerability which may not fall within the community setting. It requires district actors and officials to be involved in analysing patterns of vulnerability, and to consider what they should do about it. They can do this better if informed by the analysis from the community. However information channelled through involvement of district actors should not be allowed to dominate communities' own analysis. PVA provides a holistic system to take voices of marginalised people to other levels e.g. during disasters for the release of national and/or international funds.

3) *National/international level:* a holistic analysis of vulnerability requires national actors to see the implications of their policies on the vulnerability of poor people. Risk management is not reducing one

person's vulnerability and creating vulnerability for others.

8.4.3. Conducting Participatory Vulnerability Analysis

8.4.3.1. Preparation

The preparation phase comprises awareness-raising at the country level, defining the purpose (TOR), stakeholder analysis and team preparation.

Country programme level awareness raising

This stage involves liaison with departments and projects at country level to raise awareness and discussions about whether there is need for external support and where this support will be sought. In every region or sub-region there will be people who have received training on PVA. So you may want to contact them to learn from experiences of other country programmes, or to ask for their support for your PVA exercise if it is necessary.

Purpose of conducting PVA

This stage involves defining the terms of reference for the PVA exercise, i.e. why is PVA being conducted? It also involves analysing background information.

a) *Developing TORs:* The information you will need to collect, the depth of the analysis, the time you need to conduct the PVA and the amount of funds necessary will largely depend on the reasons for conducting a PVA exercise. This will have to be agreed with all parties, including the community. However a PVA may be conducted:

    1) to diagnose vulnerability as well as its causes (this may be done as a baseline that takes a broad view of vulnerable situations)

2) to focus on specific vulnerable groups, hazards or locations or

3) to inform better emergency preparedness, mitigation and response as well as better development work (this may be for a new or existing programme or overall strategy).

b) *Analysing secondary data/background information:* analysis of existing information relevant to the objective and study area is part of the process of defining the purpose. This is necessary to avoid researching information that is readily available. It is however important to make a judgement on the validity of existing secondary information – in many developing countries such information may be outdated or may not exist at the lower administrative levels. This activity gives ideas for information gaps and finding ways of bridging those gaps. A simple information gap analysis can assist in identifying the information you require to research.

- identify the information already available
- analyse the information available based on steps for analysing vulnerability
- identify the information required but not yet available.

The matrix allows you to cluster the information against the PVA analytical framework.

Stakeholder analysis

This activity will help you identify the range of stakeholders that may be involved in the PVA process. It also suggests methods for working out how to involve them, at what stage etc.

- Identification of stakeholders to be involved in the PVA exercise depends on the purpose. The success of the PVA is partly in the diversity of stakeholders involved. This includes the communities in areas where PVA will be conducted. Communities would need to feel ownership, otherwise they may feel that the exercise is externally driven.

— Agree with stakeholders on the aims and objectives of the exercise, as stipulated in TORs.
— Make logistical arrangements. Think of what will happen and when.
— Plan district/national level feedbacks: venue, facilitation, who will be invited, what outputs will be required.

Team preparation

The aim of team preparation is to ensure common understanding of both the field exercise and the PVA process among team members. A preparatory workshop will assist the team to prepare well. This workshop may take 2-3 days. Possible topics to be covered may include the following:

— Reaffirming the purpose of the PVA being planned.
— Understanding vulnerability and PVA –this may include translating key words into local language. It helps the team also to understand the difference between poverty and vulnerability.
— Simulation exercise for PVA step-by-step analytical framework.
— Developing a plan of action for the fieldwork.
— Assigning roles and responsibilities.
— Logistical arrangements.
— Ensuring that communities are informed well in advance.

Phase 2: The analytical framework

The analytical framework uses a step-by-step guide. The steps include:

Step 1 — situation analysis of vulnerability

Step 2 — analysis of causes of vulnerability

Step 3 — analysis of community action

Step 4 — drawing action from analysis.

When analysing vulnerability it is important to note that it is a predictive analysis: in other words, it is an analysis of people's conditions and how these may predispose them to harm. The analysis breaks down vulnerability into detail to the point that it can be understood by both communities and the actors who support them.

Phase 3: The multi-levelled analytical approach

Analysis of vulnerability is conducted at three levels: community level analysis, district level analysis and national level analysis.

Community level analysis

The aim of this section is to outline activities to be done at the community level for methodically analysing information relating to vulnerability, hazards, risks and capacities. The field level analysis involves:

1) Conducting community level analysis –using the step-by-step analytical framework (outlined above), conduct discussions with communities in the areas you have selected. Community discussions may take 3-4 days.
2) Ask the community to select representatives for district level analysis and feedback processes.
3) Compile the data sourced out from the discussions into meaningful matrices.

4) Prepare for district level analysis and feedback, as a team. You can do this by:
   - isolating key issues relating to causes of vulnerability, coping mechanisms, external support used to reduce vulnerability
   - pulling out community level actions, district level actions and national level actions developed by the community
   - agreeing on roles and responsibilities e.g. assign someone to make a presentation, another to take notes, another to facilitate the discussions
   - assigning some roles to the community members who will be attending the feedback process.

District level analysis

District level analysis is about using the PVA findings to make change. It ensures that further action systematically applies the research data to build on already established processes. It is at this level that sharing knowledge with all stakeholders (particularly secondary stakeholders who may not have been directly involved in the research) is very important. This is an opportunity to track down information flows and begin to challenge the decision-making process. This can take the form of a one-day workshop following the pointers below:

1) Start the discussion by asking for their perceptions (on vulnerability) using the step-by-step analytical framework. Participants can work in groups. Explore existing measures to protect people or reduce vulnerability to various shocks and hazards.
2) Make presentations of key issues emerging from the field and action plans (from the community analysis).
3) Plenary – get feedback on the community level analysis

4) Discuss policies in place, and how well they are implemented; how does vulnerability (as presented from the community analysis) relate to these policies, people's rights and legal frameworks? For example, what provisions are made in the district to reduce vulnerability of rich people and property while neglecting or exacerbating the vulnerability of poor people? (e.g. embankments/fence protecting rich people's land/animals; drains moving water out of developed areas).
5) Ask the participants to agree on actions!
6) Inform them about next steps – for example the national level feedback process.
7) At this stage, by recording the deliberations on audio or video, make your best report!
8) Prepare for the national level analysis and feedback.

National level analysis: is similar to the district level analysis but the difference is that at the district level, you are dealing with policy practice and at the national level you will be dealing with policy makers. This process can be used to crosscheck with government departments and other players on how they are reducing poor people's risks/vulnerability. This stage could take the form of a one-day meeting or workshop.

1) Start the discussion using the step-by-step analytical framework to get the participants' perception about vulnerability.
2) Present key issues emerging from the community and district level analysis.
3) Agree on how to feed the information into other national level processes of analysing vulnerabilities e.g. Vulnerability Assessment Committees (VAC).
4) Agree on next steps.

5) Document the proceedings by video or report.
6) You may want to write a report for the process.

International level feedback: involves linking issues on vulnerability reduction from the community, district and national level to the international level. This may involve:

1) sharing reports through networks e.g. prevention, Southern Africa Vulnerability Initiative (SAVI)
2) working with ActionAid International's emergencies team or other ActionAid International advocacy groups on particular advocacy themes
3) using the process to inform vulnerability analysis at the international level.

9

Desertific Hazard Assessment

9.1. Influencing Factors of Desertification Hazards

There are several of the factors affecting the processes of desertification. These, of course, depend on the great variety of physical and land-use characteristics of an area. Only those physical characteristics that can be easily measured or calculated, and which do not vary greatly with different land-use practices are considered to be principal factors; they are precipitation, potential evapotranspiration, soil texture, land form, and wind. Because man can cause, intensify, or ameliorate the processes of desertification through the activities of the yearly agricultural cycle, it is important to know how these activities respond to the principal physical factors.

The occurrence of soluble salts and saline conditions in the substrate, soil structure, soil nutrients, and existence and movement of herbivores, including insect fauna, are also very influential in desertification. These, however, can rapidly change under conditions of use.

9.1.1. Precipitation and the Occurrence of Drought

Data on annual precipitation levels are generally available although monthly figures are often scarce. In cases where data are missing, annual and seasonal amounts may be estimated from the observation of the types and densities of the native undisturbed vegetation. Precipitation levels may be estimated, for example, from life zone maps developed with the Holdridge life zone system (Table 1). Information concerning historical annual variation of precipitation will be needed in order to gain an insight into drought occurrence. If climatic station data are not available, secondary information may be obtained from written and spoken historical records, geomorphological studies, as well as analysis of growth rings in woody vegetation.

Table 1. Relationship between precipitation and dominant vegetation

Precipitation	*Dominant Vegetation*
0-25.4 cm/yr.	Desert
25.4 cm-76.2 cm/yr.	Grassland, Savanna, Open Woodland
76.2 cm-127.0 cm/yr.	Dry Forest
> 127.0 cm/yr.	Wet Forest

In this primer precipitation levels greater than 1500mm/year are considered to be too humid for most forms of desertification. Thus, if the study area's precipitation level is below 1500mm/year, the methods discussed here may help in the planning process.

Several different types of storms are important in the analysis of desertification hazard. Cyclonic or frontal storms are long-lived and move almost continuously in definite routes across a continent. In areas where most of this type of precipitation exists, long periods of drought can occur. Orographic precipitation is caused by rising air currents that travel over land mass at high enough altitude that expansion and cooling of the air mass causes moisture condensation. As the air mass descends after crossing a higher elevation, it is warmed and the available moisture is tightly held. This creates arid conditions on the lee side of the elevated areas. Such is the case in much of Central America, where air movement from the Caribbean Sea contributes to the formation of cloud forests on mountain peaks but extremely arid conditions at lower elevations on the western side of the range.

Convective precipitation occurs in hot months when the land surface becomes heated under strong insolation which then heats the lower strata of the atmosphere causing them to rise to strata of cooler temperature. Condensation causes rains, which tend to be heavy, of short duration, local in distribution, and accompanied by lightning. These storms are often accompanied by strong winds, and, at times, only the winds occur with little or no precipitation, causing intense dust storms. Because of both orographic and connective precipitation, rainfall maps made at stations a few miles distant in mountainous country may be subject to considerable error.

9.1.2. Potential Evapotranspiration (PET)

The concept of potential evapotranspiration is defined as an estimation of evaporation and transpiration rates if soil water is not limited. It compensates quite easily for the lack of information on transpiration and allows a clear synthesis of the numerous measurements of soil moisture, infiltration, runoff, etc., that are needed to understand climatic parameters. Evapotranspiration rates are related to several climatic factors, the most important one being temperature. For example, adjusting temperature figures for variations in day length (hours of daylight) using a formula developed by Penman demonstrates that there is a close relationship between mean temperature and potential evapotranspiration. Consequently, this formula may be used to compute potential evapotranspiration for any place whose latitude is known and where temperature records are available or can be estimated. Data on water surplus and deficit can be inferred by comparing monthly precipitation and monthly potential evapotranspiration figures.

Evaporation rates can be obtained from readings on controlled bodies of open water (evaporation pans). Although transpiration is a product of evaporation from leaf surfaces, its rates depend on the availability of soil water as well as the structural and functional features of the plant (location of stomata and the internal processes governing loss and gain of water in the guard cells) as these are influenced by light. For example, light increases transpiration rates more than it does evaporation rates. On the other hand, wind increases evaporation rates more than it increases transpiration rates. Thus, evaporation rates do not always indicate transpiration rates.

9.1.3. Wind

Wind is a climatic factor that can intensify desertification in many ways. Its force can erode, transport, and deposit soil particles. Damage to plants can occur either through the impact of its physical force when velocities are high or through the impact of transported abrasive soil and salt particles (sand blasting). In dry areas where soil is not held in place by vegetation, wind is a major factor in the formation of dunes. In the formation process of these dunes the wind, due to its velocity, leaves coarser material behind and continues transporting the finer soil particles. Although dunes can exist in non-desertification environments like coastlines or close to loosely cemented sandstones, their movement towards the outside limits of deserts is a clear indicator that desertification is taking place.

Wind increases water evaporation rates from land and plant surfaces. This evaporative power of moving air increases with higher temperatures and decreased relative humidity. As a result, hot dry winds during a plant's growth period can increase the amount of water it uses. Although wind is a part of the climate and is much more regional in scope, wind patterns can change drastically under the influence of man's activities through the removal or addition of vegetation-especially woody vegetation-which acts as a barrier, provides shade, and decreases albedo.

9.1.4. Soil Texture

Soil texture can influence many other soil characteristics, especially those concerned with soil moisture and soil fertility. Sandy soils that are irrigated require relatively more water than finer textured soils but an excess of water may leach away any available colloids and nutrients. On the other hand, because precipitation penetrates almost immediately on coarse-textured soils, runoff is reduced to almost nothing. Fine-textured soils may hold more water than coarse-textured soils but in general: (a) they hold it in the upper soil layers where drying is greater; (b) there is greater water loss due to lower rates of infiltration and higher rates of overland flow or runoff; (c) they restrict root growth and seedlings, which may sprout on such soils and die before reaching the moisture held at deeper soil levels; (d) they are responsible for shallow root growth, which makes the plant susceptible to drought; and (e) they are less susceptible to gully and sheet erosion.

9.1.5. Land Form

Two land form characteristics of desertification are: (a) degree or steepness of slope and (b) depth from soil surface to the water table.

Steepness of slope is important because it influences the velocity and the amount of surface water flow. Runoff, of course, is greater if the hillside slope is steeper. Slope steepness also influences amount and intensity of sunlight that a particular site receives. Desiccation, or drying, is greater if the slope faces the sun for longer periods of time and increases further if the slope angle is perpendicular to the sun's rays. Due to the erosion agent running water, particles are carried to flatter areas or areas of depression. Thus, soils tend to be shallower and of coarser texture near the top of a hill and are relatively deeper and finer textured at the foot of a slope. Desiccation is less severe in areas that face the sun less often and are therefore largely in shadow, again depending on the steepness of the slope. Desiccation is also lower in areas with leeward positions that are protected by intervening lands of higher elevations. If such elevations are high enough and abrupt enough, a rain-shadow effect could be established that would cause a decrease in overall precipitation.

Depth to ground water is important because if the water table lies too deep, the plant roots will not be able to obtain the available moisture. On the other hand, if the water table is too close to the surface, waterlogging will become a problem. And, in these areas, saline and alkaline conditions can kill vegetation or decrease their growth rates.

9.1.6. Land Use

How a landscape is used can initiate the process of desertification. Certain kinds of agricultural practices, overgrazing by livestock and wildlife, extractive forestry, construction activities, and the use of fire are often considered the important contributors to the process. Dry-land agricultural practices can contribute to the process because they expose soil to wind and water erosion during periods of early planting and after harvest. The finer soil particles are blown or washed away with the essential organic matter that will be missed in the following agricultural cycle. Thus, a gradual reduction of nutrients occurs through the years.

Irrigated agriculture also may contribute to desertification if it is responsible for waterlogging and salinisation. Waterlogging reduces soil aeration and the plant roots are unable to survive in this soil. This condition worsens the closer the water table is to the surface. Salinisation or alkalisation of lowland areas occurs when excess irrigation induces accumulation of soluble salts, which also impairs plant growth.

Grazing by domestic livestock, feral or exotic animals, and both large and small game animals, if poorly managed, contributes to loss of vegetative cover for the soil. In some ecosystems overgrazing promotes invasion by woody plant species that the grazing animals find unpalatable. Thus, the biomass level increases with a less desirable mixture of plant species. Competition for available soil water between the plants combined with the continuing overuse of the palatable species by the grazing animals can cause the rangeland to deteriorate further in terms of its production of fodder and animals.

Cutting of firewood for both domestic and industrial purposes likewise can contribute to desertification. Firewood collection, and charcoal production, normally become significant in areas near population centers where this is the cheapest or only source of energy. In other areas the collection of firewood on recently logged or burned sites is of secondary importance. Collection for industrial purposes can rapidly and significantly reduce vegetative cover since the demand is high and the gatherer obtains income from collecting wood.

Like agriculture, construction of buildings, reservoirs, roads, etc., and indiscriminate use of fire also remove the vegetative cover and leaves the soil unprotected and susceptible to erosion. Activities such as these, which change the normal drainage patterns, can be responsible for the erosion of extremely large amounts of soil. Almost any disturbance of stable soil surfaces such as desert pavement can initiate a new cycle of wind and water erosion.

9.1.7. Land Management

The consequences of land management practices can be positive or negative. It is estimated that 23 million metric tons of wheat production a year are lost throughout the world to desertification. Through proper agricultural, forestry, and rangeland management techniques many of these losses can be minimised. Enriching soils whose nutrients have been lost is costly and can be prevented.

Management of land, however, involves much more than land itself, and must consider other physical, biotic, social, economic, and cultural attributes. In many parts of the world these attributes lead to an annual cycle of events that represent what can be called, in a rural context at least, the agricultural year. In such areas, a year can be divided into distinct periods that depend on the number of crop cycles that can be grown in one year's time. This number is related to the length of the growing season, which may be dictated by temperature and day length-photoperiod-and by how precipitation is distributed over the year. All of these factors, of course, influence plant growth, flowering, and seeding and thus dictate the kinds and timing of activities that a farmer, forester, or livestock producer will undertake during the year.

In temperate climates there are generally four distinct periods which dictate the agricultural year. Given adequate and evenly distributed precipitation, the agricultural year follows this cycle: spring for land preparation and planting; summer for growth of the crop and the activities of cultivation (weeding, fertilising, etc.); fall for the activities of harvest; and winter for fallow. These will vary greatly as one moves toward more tropical climates, where more than one crop cycle may be possible, sometimes with almost no fallow period. It also varies as one moves toward more arid climates, where irrigation may be necessary to replace natural supplies of moisture or where the fallow period may be quite lengthy because of the lack of precipitation.

Over time, a cycle of activities evolves that fits the climatic pattern for a given site. Problems arise when drought throws the cycle out of phase and wind or heavy rain occurs when the vegetative ground cover has been removed or disturbed. Problems also occur in areas that have been recently opened to cultivation, and where a proper crop/climate fit has not been developed; or where new crops have been introduced that do not quite match the peculiarities of a local climate. The problem of soil erosion arises when land lies in a state of preparation or uncovered fallow during periods of wind and heavy rain or if droughts follow land preparation and seeding.

Management of livestock, especially ruminants, must also fit the local climatic and biotic cycles. Heavy grazing in spring, for example, when grass is young and the ground is wet, can cause problems of trampling and soil compaction, while excessive grazing pressure during periods of drought can uproot plants and place even more stress on vegetation that is struggling to survive and reproduce.

Many other variations and combinations occur in the myriad climates that exist in arid and semi-arid zones regardless of whether they are tropical or temperate. The activities of agriculture and livestock management should be matched against the agricultural year to evaluate if moisture deficits, wind, and bare surfaces occur and if they occur together.

9.1.8. Potential Desertification

There are two important aspects of the identification and evaluation of desertification potential: the zoning of hazard potential and the identification of the specific desertification hazards.

9.1.8.1. Hazard zoning

The study area in which the desertification hazard evaluation is to be carried out should be zoned according to as many of the major variables active in the desertification process as possible. Climate data are especially important and a number of conventional systems of climate zoning exist. Bailey suggests several such classifications depending on the level of mapping being considered. For example, the methods of Koppen, Kuchler, and Hammond can be used to divide the landscape into a series of ecosystem levels. The Holdridge life zone method uses measurements of mean annual "bio-temperature," potential evapotranspiration ratio, and average total annual precipitation to divide an area into life zones and has a number of advantages. First, it is based on biotemperature and moisture-both highly correlated with desertification. Second, it now includes data on slope, soil texture, and soil depth. And, third, maps at varying scales exist for most of Latin America ar some of the Caribbean.

9.1.9. Identification of Desertification Hazards

Desertification is a complex phenomenon requiring the expertise of specialists from several different disciplines if it is to be understood and managed. Climatology and meteorology, soils, agronomy, range management, anthropology, political science, and economics are all appropriate for undertaking a study of the desertification process, and input from these and others will be required if development planning is to adequately treat the subject.

10

Forecasting and Warning of Disasters

The statistics of recorded disaster data show that over the last decade (1995-2004) nearly 6000 disasters were recorded, accounting for about 900,000 dead, US$ 738 billion material losses and 2500 million people affected. Disasters have mostly hydro-meteorological origins, from extremes of wind, rainfall and temperature, but earthquakes figure high in the death rates, owing mostly to inadequate building design. Disasters disproportionately affect poor people and poor countries and are increasingly recognised as a major handicap to the development of many countries.

Figure 1 shows the rising trend in the number of people affected by disasters over the last 35 years. Disaster impacts are generally increasing as a result of the combination of increasing populations, greater concentrations of people and assets in vulnerable areas, greater use of insurance and the modification and degradation of natural environments, such as floodplain settlement, coastal exploitation, wetland destruction, river channelling, deforestation, soil erosion and fertility decline. Vulnerability to hazards is exacerbated by poverty, disease, conflict and population displacement.

The estimation of long-term trends in disasters depends somewhat on the period used and the dataset.

Comparing the most recent decade 1995-2004 with the previous decade 1985-1994, the CRED data shows the number of people affected increased 1.5 times, economic damage increased 1.8 times and total deaths increased 2.0 times. The latter figure is heavily affected by the 26 December 2004 tsunami tragedy. Prior to that date, the trend in death rates since the 1950s was downward, as a result of improving early warning systems, better preparedness and response, including systematic food aid systems. Together these now avoid the massive famines and flood losses that earlier prevailed.

10.1. Elements of Early Warning Systems

The expression 'early warning' is used in many fields to mean the provision of information on an emerging dangerous circumstance where that information can enable action in advance to reduce the risks involved. Early warning systems exist for natural geophysical and biological hazards, complex socio-political emergencies, industrial hazards, personal health risks and many other related risks. In the present setting we are concerned with geophysical hazards-storms, floods, droughts, landslides, volcanic eruptions, tsunamis, etc-and related hazards that have a geophysical component, such as wild-land fire, locust plagues and famines.

The concerns of early warning researchers and practitioners therefore span the natural and social sciences and theoretical and practical matters. To be effective and complete, an early warning system needs to comprise four interacting elements, namely: (i) risk knowledge, (ii) monitoring and warning service, (iii) dissemination and communication and (iv) response capability. While this set of four elements appears to have a logical sequence, in fact each element has direct two-way linkages and interactions with each of the other elements.

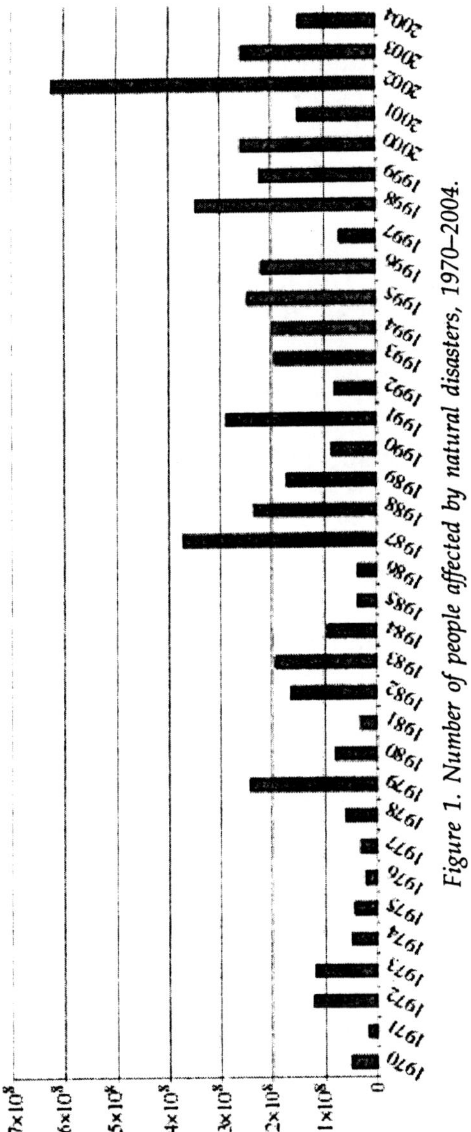

Figure 1. Number of people affected by natural disasters, 1970–2004.

The second element, the monitoring and warning service, is the most wellrecognised part of the early warning system, but experience has shown that technically high-quality predictions by themselves are insufficient to achieve the desired reduction in losses and impacts. The human factor in early warning systems is very significant. Failures in early warning systems typically occur in the communication and preparedness elements.

This was true of Hurricane Katrina which affected New Orleans in late August 2005, though in this case there was the additional failure in respect to risk knowledge, namely a lack of full public and political appreciation of the core vulnerability of the inadequate levees and the consequences of their structural failure or overtopping by storm surges. In the case of the December 2004 Indian Ocean tsunami, there were major failures in all four elements.

It should be noted that in order to sustain the four elements over the long run, it is necessary to have strong political commitment and durable institutional capacities, which in turn depend on public awareness and an appreciation of the benefits of effective warning systems. Public awareness and support is often high immediately after a major disaster event-such moments can be capitalised on to strengthen and secure the sustainability of early warning systems.

10.2. Methodologies for Monitoring and Forecasting

If disasters arise from the concatenation of multiple factors, natural and social, then in principle at least, an early warning system should address all of the factors relevant to the particular risk. From this perspective it is desirable to monitor and provide early warning and foresight not only on the short-term precipitating hazards and geophysical conditions but also on th[illegible]evant longer-term factors such as declining environm[illegible] state, risk-raising development practices and projects, risk-altering policy changes, the status of social communications and capacities, trends in food markets, settlement trends and migration, conflict and health status. This involves a wide range of time frames, as illustrated in table 1, and diverse methodologies for monitoring and forecasting.

Table 1. Illustration of factors of relevance to early warning systems and their time frames in seconds (S), minutes (M), days (D), weeks (W), months (M), years (Y) and decades (D).

factor	*time frame*						
	S	M	D	W	M	Y	D
Seismicity, tsunami	X	X	X				
Weather, oceans, floods		X	X	X	X		
Soils, reservoirs, snow pack, El Nin				X	X	X	
People exposed, conflict, migration			X	X	X	X	
Crop production, prices, reserves, food aid				X	X	X	
Environmental management and state					X	X	X
Industry, urban, infrastructure design					X	X	X
Land use planning, climate change						X	X

Extending this line of thinking, one can argue that the citizen and the public risk manager is not so concerned with the specifics of particular hazards, but rather the package of risks faced and how to mitigate and prepare for them. This implies that an approach that addresses all relevant hazards in an integrated fashion, and not as separate unconnected systems, is more appropriate to the management of natural risks. Such a 'multi-hazard' or 'all-hazard' approach should provide synergies and cost-efficiencies, e.g. in data gathering and processing and in public preparedness efforts, and should assist in sustaining warning capabilities for the more infrequent hazards, such as tsunamis.

It is important, however, not to gloss over the very specific characteristics of the different hazards. For example, tsunamis and storm surges both cause coastal inundation but the detection and monitoring methods, lead-time, duration of the hazard and response actions are very different. A multi-hazard approach should not be allowed to force generalities or centralised control upon warning systems, but must be tailored to the needs of each hazard and built upon the specific technical capabilities required and the available institutional capacities. The need is for a coordinated 'system of systems'. Much remains to be elaborated in the practical implementation of these ideas.

10.2.1. Linear Paradigm of Model-based Early Warning Systems

The most common current view of early warning systems comprises a 'warning chain', a linear set of connections from observations through warning generation and transmittal to users. In the meteorological community the term 'end-to-end' warning system is often used. The end-to-end concept aims to make forecasts and warnings more relevant and useable to end-users, and has evolved partly in response to the commercialisation imperative in many national meteorological services, as well as through efforts to make better practical use of the probabilistic and weakly predictive seasonal forecasts of the ElNino phenomenon. It emphasizes the necessity to have all the links in the early warning chain in place and systematically connected.

At the heart of all early warning systems is some sort of model that describes the relevant features of the hazard phenomenon and its impacts, particularly their time evolution. The model provides the means to make projections of what might happen in the future-and therefore what actions might be desirable in response.

Models may be as elaborate as the physics-based global numerical weather prediction models, or as straightforward as 'common knowledge' mental models (e.g. that the noisy approaching tsunami wave will arrive in a few minutes). They may be slowly evolving, as in a drought model where the loss of soil moisture may occur over months, or very rapid, such as in an earthquake where the differential speed of electromagnetic signals relative to seismic waves can be used to automatically shut down a distant sensitive system a few seconds before damaging stresses occur.

Models also underlie the other parts of the warning system, such as the likely impacts of a hazard, the way warnings are communicated and acted on, and the dynamics of evacuation processes, but these vulnerability and response process models are generally much less developed than the geophysical process models.

All models are driven by a specification of an initial state, which must be obtained by observations (or from the output of an upstream observation-driven model). Observation systems can be expensive to install and operate and are often rather inadequate, especially in poorer countries. The initial state is, therefore, always imperfectly known, owing to imperfect spatial representation, instrument error and absence of data on some relevant factors.

These uncertainties of the initial state propagate through the models, and together with errors in the model physics and representations thereof and random noise factors, result in uncertainty in the model estimates of future conditions. Warnings are, therefore, inherently probabilistic, even if based on sound physics and presented in a categorical format. Of note are forecasts of seasonal climate anomalies, which are strongly affected by system noise and uncertainty, and can only be represented in probability terms, and where it must be left to the end-user to judge the possible impact consequences of the projected possible climate outcomes.

Currently, tsunami warnings mostly are based on simple statistical relationships with precursor seismic observations, but these latter observations do not allow accurate prediction of the oceanic response, and so the false warning rates are high and the probability characteristics are poorly known. Usually, the warnings are provided only in categorical forms that usually require immediate response action. However, developments in ocean observation systems and in ocean wave propagation and coastal inundation models are in place to improve this situation in the near future.

10.2.2. Limitations of the Linear Paradigm

Scientists and technologists are typically the core stakeholders in early warning systems, as they are the custodians of the geophysical and technical knowledge base upon which the warning system relies, and they are generally very motivated to use that knowledge for the good of society. As a result, early warning systems tend to be largely conceived as hazard-focused, linear, topdown, expert driven systems, with little or no engagement of end-users or their representatives. It can be noted, however, that people generally are not interested in early warning systems until some personally threatening event arises, and so most of the time are happy to leave the matter to the experts. While the prevailing end-to-end linear paradigm is an advance on previous techno-centric concepts it nevertheless retains a number of shortcomings, as follows:

(i) the focus still tends to remain on the hazard, with less emphasis on the vulnerabilities, risks and response capacities,

(ii) the different hazards are typically dealt, with by separate independent technical institutions, with few synergies or mutual benefits being sought,

(iii) the dominance of the expert can lead to difficulties in user appreciation of such things as the meaning of a warning, warning uncertainty, the nature of false alarms and the necessary responses to different types of warnings,

(iv) the role of research and knowledge from outside the core area of expertise is often not acknowledged,

(v) there is little engagement or empowerment of those at risk in the design and operation of the warning system, and hence a tendency by users to lack any sense of ownership in the system and to mistrust the experts and authorities,

(vi) there are few systematic mechanisms to improve the system through the incorporation of the knowledge, experience and feedback from users and those at risk, and

(vii) weak public engagement and recognition tends to lead to weak political and budgetary support for the warning system.

The Hurricane Katrina disaster is a case in point where the meteorological warnings of wind speed, storm surge and rainfall were accurate and frequently communicated many hours in advance but the public and official engagement and responses to the warnings were inadequate. Similar experiences elsewhere have shown that to be effective, early warning systems must be both technically systematic and people-centred.

The 'people-centred' characteristic requires many systematic approaches and diverse activities spanning the four elements of early warning systems described above, such as: identifying target populations, especially the vulnerable and disadvantaged and interacting with them to determine needs and capacities; conducting town meetings and involving communities in exploring and mapping their risks and planning their responses; fostering the development by communities of monitoring and warning systems for local risks; generating public information tailored to target groups and making innovative use of the media and education systems; establishing people-focused benchmarks and performance standards for technical warning services; developing formal mechanisms for public representatives to monitor and oversee warning system design; using surveys to measure public awareness and satisfaction; creating monuments, publications, annual events and other anchors of public memory and learning; providing training on social factors for technical experts, authorities and communicators who operate the warning system; conducting research on factors that enhance or impede human understanding of and response to warnings; and providing exercises and simulations to enable people to experience and practice warning interpretation and responses.

It is important to recognise that these diverse activities cannot be undertaken or directed by any one organisation, but require the coordinated participation of many different types of organisations, bound by a consensus of commitment to the 'people-centred' concept, and to the idea of an integrated system that is measured by its performance-namely protecting those at risk. National platforms for disaster reduction, stakeholder roundtables or inter-departmental committees should be empowered or established to organise the required coordination. The core technical agencies can play a key role by demanding the establishment of such mechanisms and supporting them with specialised technical information.

10.2.3. Integrated Systems Model for Early Warning Systems

Early warning systems have evolved in line with the development and application of scientific knowledge. Four developmental stages can be distinguished:

(i) pre-science early warning systems. Warnings, if any, may be based on unrelated factors such as meteor occurrence, cloud shapes, plant flowering or fruiting performance, etc., but also may be based on indigenous observations of relevant factors such as the state of the oceans or visibility of the stars,

(ii) ad hoc science-based early warning systems. These are systems such as are often established on the initiative of scientists or community groups concerned with particular hazards, such as near-Earth space objects, a nearby volcano or a flood-prone river,

(iii) systematic end-to-end early warning systems. The best known and most developed are those of national meteorological services, for weather-related hazards. Typically these systems operate under a country-wide mandate and involve the organised, linear and largely uni-directional delivery by experts of warning products to users, and

(iv) integrated early warning systems. This concept, as proposed here and illustrated in figure 2, emphasizes the following characteristics: the linkages and interactions among all the elements necessary to effective early warning and response, the role of the human elements of the system and the management of risks rather than just warning of hazards.

The integrated model proposed in figure 2 includes the core warning system elements, but in addition contains two new key features. The first is the inclusion of actors that often are not recognised as part of the warning system, most notably the political-administrative supporting entities, the district and community actors and the research community.

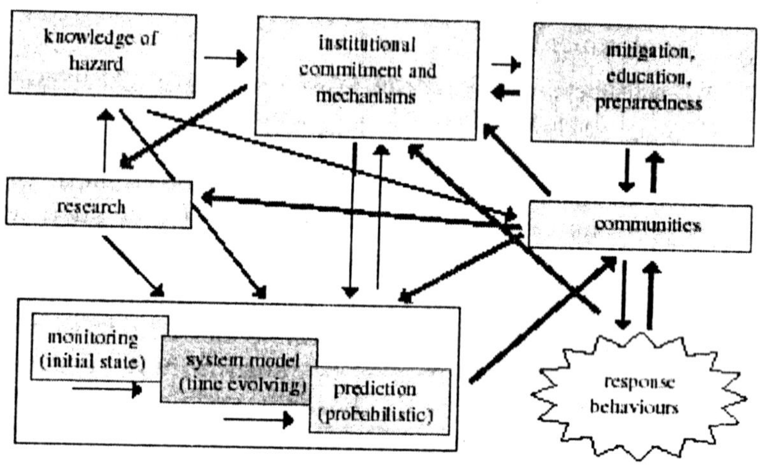

Figure 2. Integrated systems model of early warning system.

The second feature is the explicit inclusion of multiple linkages and feedback paths, particularly from affected populations through their organisations to the political and technical actors. The model could be elaborated further for the particular circumstances of countries, e.g. to better specify the district-level and community-level elements or the collaborative roles of different discipline-based technical institutions (e.g. such as seismological, oceanographic and meteorological organisations in a tsunami early warning system).

Figure 2 is largely conceived as a nationally based system, but it is worth noting that many warning systems depend on regional and international cooperation to secure the exchange of necessary data and warnings. This is not a simple matter to arrange, however, as sovereign states can view their data as having strategic or commercial value, and for these reasons can deny or limit its exchange. In the field of meteorology, many years of discussion under the auspices of the World Meteorological Organisation (WMO), a specialised technical agency of the United Nations, have led to formal agreements on the types of data that are routinely exchanged. Much remains to be done to achieve similar levels of agreement in other hazard fields, e.g. in respect to rainfall and river flow data required for flood warnings in shared river basins and seismic data for tsunami warnings.

Underlying the integrated model is the important foundational assumption that we are dealing with a system, defined here as a set of elements and associated linkages designed to achieve a particular result-namely the reduction of risk for target populations and assets through early warning. The system is judged on its effectiveness at delivering the desired result, and can only be effective if the elements and the linkages are well-understood, well-designed and well-operated.

10.2.4. Systems-oriented Research needs for Early Warning

Early warning systems require a broad multidisciplinary knowledge base, building on the substantial existing discipline-based research in the geophysical, environmental and social science fields.

10.3. Recent Develop to Better Early Warning Systems

The December 2004 tsunami shone an intense spotlight on questions of early warning systems and preparedness, leading most notably to the call by United Nations Secretary General Koffi Annan in January 2005 for a global warning system for all hazards with no country left out. This was to be followed later in the year by his request to the International Strategy for Disaster Reduction (ISDR) secretariat to coordinate a global survey of early warning systems, with a view to identifying gaps and opportunities, as a basis for developing such global capacities.

Meanwhile, negotiations by states over 2004 culminated in a major international agreement on disaster risk reduction at the World Conference on Disaster Reduction in Kobe, Japan, 18-22 January 2005, namely the Hyogo Framework for Action 2005-2015: building the resilience of nations and communities to disasters. The topic of risk and early warning is one of its five priority areas for action.

Leading UN agencies announced at the conference the launch of an International Early Warning Programme (IEWP), as a vehicle to stimulate and coordinate cooperative initiatives to advance early warning methodology and to build early warning capacities. Shortly afterwards, Germany offered to host a third International Conference on Early Warning (EWC III) under UN auspices.

Rapid progress has been made on developing a tsunami warning system for the Indian Ocean, with strong support by the countries affected and by the international donor community, including through a multi-partner, multi-donor US $11 million project coordinated by the ISDR secretariat.

This project has underwritten the important work of UNESCO's Intergovernmental Oceanographic Commission to upgrade regional seismic and oceanic observation systems, to assess national technical needs and to establish intergovernmental coordination mechanisms. It has also supported WMO efforts toward upgrading meteorological telecommunications networks to handle high-speed tsunami information transfers, as well as projects by United Nations organisations and

Asian regional disaster organisations to improve public awareness and disaster preparedness.

The project seeks to link and integrate these various initiatives into a strategy to build long-term disaster risk reduction and risk management networks and policies. Separately, the IOC is building the necessary global institutional framework to support tsunami early warning systems in other at-risk regions such as the Mediterranean, Caribbean and Central America.

In early 2005, the British Government established a Natural Hazard Working Group under the guidance of the government chief scientist to advise on the mechanisms that could be established for the detection and early warning of global physical natural hazards, particularly those hazards that could have high global or regional impact, and including international mechanisms needed to enable the international science community to advise governments.

The Working Group recommended the establishment of an International Science Panel for Natural Hazard Assessment, within the UN disaster management framework, to enable the scientific community to provide authoritative information on potential natural hazards likely to have high global or regional impact, by addressing gaps in knowledge and advising on potential future threats and on how science and technology can be used to mitigate threats and reduce vulnerability.

The Working Group also noted that the wellestablished WMO international system operated by national meteorological services for weather data gathering and warning provision provided a potential basis for strengthening other less-developed hazard warning systems.

The Working Group's recommendations subsequently were taken up in part by the 2005 meeting of the G8 ministers, who noted that 'early warning systems for global geophysical events should be based on high quality and appropriate scientific advice that can be translated into effective action by policy makers and those most at risk at a local level'. While not explicitly referring to the British proposal for a new panel, the G8 stated 'We will support closer co-ordination on natural hazard assessment to enable the scientific community to advise on potential natural hazards likely to have high global or regional impact.'

11

Rehabilitation and Reconstruction

Underdevelopment and ineffective or inappropriate development programmes increase vulnerability to hazards, and hence lead to more disasters, great and small. In turn, emergencies make subsequent development more difficult for disaster-affected communities that have lost their livelihood assets - and therefore for the institutions that are trying to help them. There is widespread agreement on the need for closer integration of relief, rehabilitation and development, which implies a longer-term perspective behind post-disaster action.

In essence, it means that relief and rehabilitation should contribute to long-term development and the reduction of vulnerability, where they can - they should not simply reconstruct the existing risk. At one time, the phrase 'relief-development continuum' was used to refer to this integration. Nowadays, it is usual to speak of 'developmental relief', a concept first articulated by the Red Cross in the mid-1990s. This expresses a broad-based and sustainable approach to post-disaster work. ActionAid uses the notion of 'recovery plus', meaning an intervention 'whereby people are in some way better off than before the emergency'.

There is plenty of scope for academics to debate the merits or drawbacks of such terms and concepts.

Operationally, it is helpful to look for the similarities in their basic principles, which can be summed up as follows:

— intervene at the earliest possible stage in the disaster cycle to protect livelihoods and reduce vulnerability;
— incorporate development principles into disaster relief operations (e. g. build up local capacities, adopt participatory approaches);
— use disaster relief not just to meet immediate needs but also to restore livelihood assets and rebuild livelihoods;
— use disaster relief to develop infrastructure that will be of value after the emergency is over; and
— take the opportunity to induce positive socio-economic change and not merely a return to the status quo.

This shift puts more emphasis on what is normally called rehabilitation. Originally seen as a distinct linking phase between relief and development, it is now seen more as a continuing process that may take place alongside both disaster relief and development, and ideally is integrated with them.

It is clearly unrealistic to expect 'normal' development to resume soon after the crisis period. People affected by disasters are left more vulnerable than they were before. Relief and development programmes need to adjust to take account of this. For example, the Centre for Sustainable Agriculture and Appropriate Technology, which works with peasant farmers in the Dominican Republic, had to redirect much of its efforts away from long-term rural development and towards short-term agricultural rehabilitation for two years after Hurricane Georges in 1998. Although interest in monitoring and evaluating relief programmes has grown enormously over the past few years, rehabilitation has not been studied as much.

Evidence of the long-term impact of rehabilitation is in particularly short supply. Where this is investigated, it is usually to demonstrate underachievement, as in the Times of India's assessment of the Orissa state government's work after the two cyclones in 1999, which found that only 11 out of 100 planned cyclone shelters had been built, only 392 out of 3, 779 secondary schools had been rebuilt, and government departments had not spent funds allocated for rehabilitation.

Many post-disaster assistance projects come to an end too soon and too suddenly. Disaster response organisations talk a lot about 'exit strategies', but what an external agency describes as a phase-out may be seen by the affected community as a cop-out, in which the agency concerned walks away rather than seeing the job through to the end. Reflecting on the experiences of Afro-Honduran communities after Hurricane Mitch, a local NGO complained of a disaster training programme established by a major international NGO that existed here for a few months, but made no provision for follow up, provided no ongoing funds for replication to expand the number of people trained beyond the initial group, and left no materials or resources for implementation of what people had learned.

Organisations that are only working in the short to medium terms in a disaster-affected area should plan their withdrawal carefully, recognising that there will be plenty of work left unfinished, and community expectations may not have been fulfilled. Phased withdrawal is preferable to sudden departure. There must be a coherent handover to locally-based organisations and communities. The process should be planned early, it should be transparent and it should be agreed with partners.

Relief and rehabilitation agencies bear some of the responsibility for ensuring that activities are sustained, and that local agents have the capacity and resources to manage this. Some relief and rehabilitation initiatives lead to longer-term risk reduction projects, especially where the same agencies are involved in both relief and development work in the area concerned. It is impossible to say how widespread this is: again, it is probably not common, but it may have become slightly more so since the spate of major natural disasters in the late 1990s.

Relief and post-relief initiatives tend to operate on different scales, with mass coverage being more easily achieved in relief operations. In the Red Sea Province of Sudan during the 1985-86 famine, a major food aid programme supported more than 400, 000 people, but a parallel rehabilitation initiative including locust control and improvements to village wells could only reach a few thousand. There is a dilemma for relief agencies in deciding how to balance the need to give relief to as many people as possible with the need to provide for future emergencies.

The shortage of funds for rehabilitation is another obstacle. Relief funding covers only short-term, often fixed periods (typically six to nine months) and so cannot be used to support many longer-lasting activities that would improve resilience. This makes no sense logically. It results from the distinction between 'relief' and 'development' in donor budget lines. It leads to effort and resources going into activities that are not sustained, and to strict limitations on activities deemed too 'developmental' by relief donors.

UK-based international NGOs interviewed in a recent study were frustrated by the inflexibility of donor regulations in this regard. Funding from development budget lines is not a realistic alternative owing to the length of time taken for major funding schemes to reach decisions. Staff in a development NGO once told the author of problems it had had with a humanitarian aid donor after a hurricane in the Caribbean. With the trees grown by its community forestry project flattened, it sought to make the best of the situation by using the wood to build shelter for people living in shanty towns, who were hurricane victims. The donor approached turned the proposal down because shelters made of wood were considered too permanent to be emergency response - but the donor was prepared to pay for bringing in plywood boards and plastic sheeting from outside the island for emergency shelter.

11.1. Approaches to Risk Reduction

11.1.1. Rebuilding Livelihoods

Preservation of livelihoods is vitally important to poor and vulnerable people, and vulnerability is closely linked to livelihood security. After a disaster, earning a living will soon be a priority for the victims. Take, for example, the village of Rampur in Nepal and its farmlands, hit by a landslide in July 1993 that claimed the lives of 18 people and more than 70 animals. The disaster occurred during a peak period for harvesting, mending terraces and planting, and so from the third day after the disaster most villagers divided their time between rehabilitating canals and farmland and managing the chaos in their homes. In the past, relief agencies often failed to appreciate how important this is.

For example, in the case of drought, interventions are often launched only after communities have begun to dispose of essential livelihood assets as the last resort in their coping strategy. Research by HelpAge International found a major discrepancy between the perspectives of older people affected by emergencies and those of aid agencies. The aid agencies did not think that earning income would be a concern for older people, but in fact it was one of their top priorities.

Relief efforts also risk undermining local markets and incomes by flooding them with goods (e. g. food aid, shelter materials) or outside labour (e. g. in housing reconstruction programmes). Given that many relief/ rehabilitation programmes are characterised by a lack of beneficiary participation in assessment and planning, there is a danger that livelihood support activities will be inappropriate. It is difficult for outside agencies to identify key livelihood issues in the chaotic and stressful conditions after a disaster, but even rapid participatory approaches can give insights into the complexity of livelihoods.

However, disaster response programmes nowadays usually recognise the need for some livelihood support. Relief/rehabilitation aid commonly includes foodor cash-for-work schemes. It is also common to provide seeds and tools for agriculture, livestock, household utensils and shelter materials. The appropriateness of such goods (e. g. are seeds suitable for local conditions and farmers?) is much debated in the literature of humanitarian relief, but the principle of helping livelihoods and not just saving lives is generally accepted. Interest in financial assistance measures is growing.

Cash-based responses to emergencies may have potential in empowering local communities economically. In the field, micro-finance institutions establish emergency loan funds to help their clients replace or repair assets. The Disaster Mitigation Institute's livelihood relief fund finances the purchase of tools, seeds and raw materials for victims of natural hazards and riots in India. The fund has supported 9, 500 people since 1998. Agencies are also looking more creatively at ways of supporting jobs. After the 2000 floods in Mozambique, initiatives by the International Labour Organisation (ILO) included rehabilitation of the central market in Chokwe and construction of three other markets, in order to help small-scale traders resume business and make goods more easily available locally.

In India, NGOs have helped artisans to continue to earn money by such measures as organising credit and supplies of raw materials, purchasing their products, and creating temporary exhibitions to help market products. There are encouraging signs of such livelihood support initiatives making a significant difference to poor people in the months after disasters. However, little is known about their long-term impact. This is a significant gap, since it can take a very long time for livelihoods to recover fully, and in many cases they never do.

Establishing sustainable small enterprises or more secure livelihoods usually takes much longer than the limited timetables of relief programmes. Potentially valuable initiatives may not be followed through. Road repairs by humanitarian agencies in Mozambique after the 2000 floods helped the local economy as well as the relief effort, but maintenance stopped when the relief phase ended.

Post-disaster conditions are special, and it is not clear how well conventional income-generating activities can work in these circumstances. Moreover, as livelihood strategies vary greatly between and within communities, livelihood-supporting programmes need to be equally varied and based on very thorough knowledge of local conditions. Local NGOs and CBOs are best placed to undertake such work - and because they are locally based, to follow up. Participatory approaches are clearly valuable here for identifying needs, setting priorities and targeting beneficiaries. They may lead to unexpected results, such as a community workshop to design a response to drought in Ethiopia that came out firmly in support of providing food, seed, fertiliser and blankets on credit instead of as hand-outs. More project evaluation and comparative research on such issues would be helpful.

11.1.2. Public Works

Cash- and food-for-work programmes are a standard device in an emergency, intended to give temporary help to disaster victims and to provide more permanent community facilities for the longer term. By supplying food or creating paid jobs they can prevent livelihood collapse. One of the most famous examples is the 1972-73 drought in Maharashtra, India, where at one point nearly five million labourers were employed on public works by the state. The income they received under the programme enabled them to buy food in the market, and by doing so helped to prevent famine. Public works activities tend to focus on construction or repair of physical structures such as roads and schools. They are often used to improve resilience to future shocks by building mitigation infrastructure such as irrigation channels, dams and other water harvesting structures, embankments, flood shelters, and measures to stabilise hillsides (terraces, gabions and afforestation).

Rehabilitation after the 1998 Bangladesh floods, for example, saw a number of cash-for-work projects building flood shelters of raised earth on common land such as school grounds and market places. Although food- and cash-for-work initiatives can help to protect livelihoods and reduce risk, success depends on good management. Threats to success include:

— *Lack of clarity about objectives.* Most schemes aim to provide income and public facilities, but in practice these two aims can be difficult to reconcile. The need to create work quickly may lead to projects of limited value, whereas it takes a long time to set up more substantial, complex initiatives because of the level of technical, managerial and other inputs required.

— Poor targeting that fails to support those most in need or creates divisions within communities by selecting some individuals and not others. There is still some debate about the best methods of selecting beneficiaries. Inadequate planning and consultation, leading to effort being wasted on mitigation structures that are not a priority for the community, or will not work.

— Lack of commitment by beneficiaries, usually because they are not participants in the project, but are treated merely as employees. This can result in poor quality of construction. It also makes it less likely that the community will continue to maintain the newly-built facilities once the food or cash payments come to an end.

11.1.3. Windows of Opportunity

Disasters are generally believed to present a 'window of opportunity' for promoting and implementing risk reduction measures, because the consequences of failing to act are so strongly implanted in the minds of those who are affected by disasters, the operational agencies that have to respond to them, and the public policy-makers who have to manage their effects. This reasonable assumption is well-demonstrated by the number and variety of mitigation initiatives introduced at all levels after major disasters in particular. For example:

— Disasters can be an opportunity to change socio-economic relationships that affect vulnerability.

— In Mozambique, ActionAid undertook an HIV/AIDS awareness campaign in camps for people displaced by the floods in 2000. The people then dispersed to their home areas, with key contact people in each place with whom the programme could work. Such coverage would not have been possible in normal conditions, where people cannot spare the time to sit together for several hours to discuss such issues. In Central America, Hurricane Mitch prompted vigorous debate about vulnerability and how to reduce it, leading to the creation of new pressure groups such as the Foro Permanente de Ciudadanas in El Salvador that sought new laws and policies for disaster prevention. Disasters can stimulate renewed thinking about the problem, which is leading to shifts in policy in some organisations. In the UK, a series of technological disasters in the 1980s led to the creation in 1991 of Disaster Action, a mutual support group and pressure group for improved disaster management.

Characteristics of the 'window of opportunity' are said to include:residents and local officials are thinking about the problem of risk, when they do not normally do so;

— the disaster may already have forced some changes (for example by destroying unsafe buildings and infrastructure);

— the community has to make decisions about recovery; and

— technical and expert advice and resources become available from government and non-government sources.

It is hard to tell how long the window will remain open, or what conditions must be met to take advantage of the opportunity. Chances of success at

— community level may be improved by:acting quickly before the fear or enthusiasm for change created by the disaster have lessened;

— basing interventions on familiar technologies and local resources as far as possible;

— concentrating on a small number of important actions, not introducing a whole portfolio of changes that dissipate efforts;

— focusing on what is achievable - communities already hit by a disaster have many urgent problems to attend to, and they will not respond if they believe the proposed mitigation measures are beyond their reach; and

— encouraging, supporting and involving communities as participants in change.

The principles of being realistic and setting priorities apply equally at organisational level, for here too momentum can easily be lost and lessons learned are soon forgotten. A further problem among organisations is that disasters may be caught up and lost in discussion of other development issues that are currently a policy priority for the organisation concerned. This happened in Nicaragua after Hurricane Mitch, when evidence of the disaster's impact was used to support arguments over alternative economic development models more than to debate measures that addressed risk reduction more directly.

The psychological impact of disaster must also be taken into account. Posttraumatic stress can be a significant influence on the way survivors, the bereaved and responders behave after disasters, yet there is scope for discussion about the nature and consequences of such stress. Overemphasis on negative responses such as post-traumatic stress disorder and unresolved grief can lead to the assumption that people affected by disasters are passive victims, whereas in fact disaster-affected communities are the main actors in disaster response. The reluctance of some practitioners and researchers to take post-traumatic stress disorder seriously may be due in part to the fear that it will reinforce stereotypes of passivity.

The experience of disasters can even stimulate survivors and the bereaved to work vigorously for better risk reduction efforts in the long term. This issue is neglected in relief and rehabilitation work in the South. This may be because psychological recovery is assumed to be 'a community function', not a task for outside agencies. Or it may simply be overlooked. It has been suggested that severe traumatic events sometimes undermine the individual and collective will to respond. It is conceivable that what aid agencies perceive as dependency syndrome (passivity brought about by the abundance of relief supplies) among disaster victims or lack of community spirit in undertaking post-disaster recovery is - at least in part - an expression of post-traumatic stress disorder. But this is complex and contested territory, and much more investigation is needed.

11.1.4. Safer Housing and Locations

Every disaster that leaves many people homeless triggers renewed interest in rebuilding homes so that they are 'safe' or 'disaster-resistant'. Shelter relief and reconstruction programmes absorb large amounts of international aid, yet very little is known about their long-term impact in making vulnerable people more secure. It is likely that such programmes have little impact, for the following reasons:

— An emphasis on technically 'safe' housing, without certainty that such housing is affordable or culturally acceptable. Large-scale programmes are particularly likely to be technology-driven and introduce new and expensive construction technologies.

— Although reconstruction programmes can provide jobs for local builders, in many cases the builders and their traditional skills are displaced by imported technologies and labour. Communities do not acquire the skills needed to extend, modify and repair the new houses.

— Where reconstruction does create local jobs, it is not clear how sustainable these new livelihood opportunities are once the programmes funded by aid agencies come to an end.

— The focus is on houses (physical structures) rather than housing (the arena of social and economic life). Homes are not seen as places of work, learning, communication and relationship-building. Houses are built without regard for how - or if - this will improve social and economic status or reduce vulnerability in its widest sense.

— Lack of community participation. Most reconstruction projects claim that they are participatory, but there is usually an element of agency propaganda in this, and the extent and nature of such participation are often hotly disputed.

In general, participatory approaches, based on local skills and appropriate technologies, offer the best chance of long-lasting success in post- and predisaster situations alike. A common response to the destruction of housing in disasters is to resettle their occupants in safer locations as the best way to defend them against future hazard events. One of the most striking examples of this came after the Maharashtra earthquake of 1993, when the state government moved 28, 000 inhabitants of 52 devastated villages to new sites. Governments are probably best placed to undertake resettlement because of the major practical challenges, but NGOs are sometimes involved. Several NGOs' responses to Hurricane Mitch in Central America in 1998 included relocation of vulnerable communities from the hillsides where their homes had been washed away by the torrential rains.

From a purely hazards point of view, relocation makes sense. Some locations - floodplains, unstable hillsides, soils likely to liquefy as a result of seismic tremors - are inherently unsafe. It is impossible, or at least extremely costly, to make communities that live in such places more secure. After a major disaster, such as an earthquake, survivors may be so traumatised and afraid of future shocks that they are very keen to move. Provision of land can also improve livelihoods where it is used to grow crops or products used in building or craft work: there are instances of this in Central America after Mitch.

However, relocation presents considerable practical challenges, notably the cost of purchasing land and providing infrastructure, and the difficulty of securing legal title. There are examples of planned relocation projects failing because the community could not obtain public land or buy private land. NGOs need to work very closely with local authorities and beneficiaries to resolve these problems. More fundamentally, the policy of resettlement overlooks the economic and other reasons that make people settle in unsafe areas in the first place.

There is sometimes a degree of compulsion in resettlement; even after disasters, people are usually reluctant to move. In Mozambique in 2000, government policy was to move communities away from areas at risk of flooding, even though their economy was based on the fertile farmland of the flood plains. Aid agencies were forbidden to give shelter materials to anyone who was not at a government-approved site, and these were often some distance from people's lands. This led to many farmers living in grass shelters on their old lands, and only returning to their new houses from time to time.

12

National Policy on Disaster Management

12.1. Introduction

India has been traditionally vulnerable to natural disasters on account of its unique geo-climatic conditions. Floods, droughts, cyclones, earthquakes and landslides have been a recurrent phenomena. About 60% of the landmass is prone to earthquakes of various intensities; over 40 million hectares is prone to floods; about 8% of the total area is prone to cyclones and 68% of the area is susceptible to drought. In the decade 1990-2000, an average of about 4344 people lost their lives and about 30 million people were affected by disasters every year. The loss in terms of private, community and public assets has been astronomical.

At the global level, there has been considerable concern over natural disasters. Even as substantial scientific and material progress is made, the loss of lives and property due to disasters has not decreased. In fact, the human toll and economic losses have mounted. It was in this background that the United Nations General Assembly, in 1989, declared the decade 1990-2000 as the International Decade for Natural Disaster Reduction with the objective to reduce loss of lives and property and restrict socio-economic damage through concerted international action, specially in developing countries.

The super cyclone in Orissa in October, 1999 and the Bhuj earthquake in Gujarat in January, 2001 underscored the need to adopt a multi dimensional endeavour involving diverse scientific, engineering, financial and social processes; the need to adopt multi disciplinary and multi sectoral approach and incorporation of risk reduction in the developmental plans and strategies.

Over the past couple of years, the Government of India have brought about a paradigm shift in the approach to disaster management. The new approach proceeds from the conviction that development cannot be sustainable unless disaster mitigation is built into the development process. Another corner stone of the approach is that mitigation has to be multi-disciplinary spanning across all sectors of development. The new policy also emanates from the belief that investments in mitigation are much more cost effective than expenditure on relief and rehabilitation.

Disaster management occupies an important place in this country's policy framework as it is the poor and the under-privileged who are worst affected on account of calamities/disasters.

The steps being taken by the Government emanate from the approach outlined above. The approach has been translated into a National Disaster Framework [a roadmap] covering institutional mechanisms, disaster prevention strategy, early warning system, disaster mitigation, preparedness and response and human resource development. The expected inputs, areas of intervention and agencies to be involved at the National, State and district levels have been identified and listed in the roadmap. This roadmap has been shared with all the State Governments and Union Territory Administrations.

Ministries and Departments of Government of India, and the State Governments/UT Administrations have been advised to develop their respective roadmaps taking the national roadmap as a broad guideline. There is, therefore, now a common strategy underpinning the action being taken by all the participating organisations/ stakeholders.

12.2. Institutional and Policy Framework

The institutional and policy mechanisms for carrying out response, relief and rehabilitation have been well-established since Independence. These mechanisms have proved to be robust and effective insofar as response, relief and rehabilitation are concerned.

At the national level, the Ministry of Home Affairs is the nodal Ministry for all matters concerning disaster management. The Central Relief Commissioner (CRC) in the Ministry of Home Affairs is the nodal officer to coordinate relief operations for natural disasters. The CRC receives information relating to forecasting/warning of a natural calamity from India Meteorological Department (IMD) or from Central Water Commission of Ministry of Water Resources on a continuing basis. The Ministries/Departments/Organisations concerned with the primary and secondary functions relating to the management of disasters include: India Meteorological Department, Central Water Commission, Ministry of Home Affairs, Ministry of Defence, Ministry of Finance, Ministry of Rural Development, Ministry of Urban Development, Department of Communications, Ministry of Health, Ministry of Water Resources, Ministry of Petroleum, Department of Agriculture & Cooperation. Ministry of Power, Department of Civil Supplies, Ministry of Railways, Ministry of Information and Broadcasting, Planning Commission, Cabinet Secretariat,

Department of Surface Transport, Ministry of Social Justice, Department of Women and Child Development, Ministry of Environment and Forest, Department of Food. Each Ministry/Department/Organisation nominate their nodal officer to the Crisis Management Group chaired by Central Relief Commissioner. The nodal officer is responsible for preparing sectoral Action Plan/ Emergency Support Function Plan for managing disasters.

12.2.1. National Crisis Management Committee (NCMC)

Cabinet Secretary, who is the highest executive officer, heads the NCMC. Secretaries of all the concerned Ministries /Departments as well as organisations are the members of the Committee The NCMC gives direction to the Crisis Management Group as deemed necessary. The Secretary, Ministry of Home Affairs is responsible for ensuring that all developments are brought to the notice of the NCMC promptly. The NCMC can give directions to any Ministry/Department/Organisation for specific action needed for meeting the crisis situation.

12.2.2. Crisis Management Group

The Central Relief Commissioner in the Ministry of Home Affairs is the Chairman of the CMG, consisting of senior officers (called nodal officers) from various concerned Ministries. The CMG's functions are to review every year contingency plans formulated by various Ministries/ Departments/Organisations in their respective sectors, measures required for dealing with a natural disasters, coordinate the activities of the Central Ministries and the State Governments in relation to disaster preparedness and relief and to obtain information from the nodal officers on measures relating to above. The CMG, in the event of a natural disaster, meets frequently to review the relief operations and extend all possible assistance required by the affected States to overcome the situation effectively. The Resident Commissioner of the affected State is also associated with such meetings.

12.2.3. Control Room (Emergency Operation Room)

An Emergency Operations Center (Control Room) exists in the nodal Ministry of Home Affairs, which functions round the clock, to assist the Central Relief Commissioner in the discharge of his duties. The activities of the Control Room include collection and transmission of information concerning natural calamity and relief, keeping close contact with governments of the affected States, interaction with other Central Ministries/Departments/ Organisations in connection with relief, maintaining records containing all relevant information relating to action points and contact points in Central Ministries etc., keeping up-to-date details of all concerned officers at the Central and State levels.

12.2.4. Contingency Action Plan

A National Contingency Action Plan (CAP) for dealing with contingencies arising in the wake of natural disasters has been formulated by the Government of India and it had been periodically updated. It facilitates the launching of relief operations without delay. The CAP identifies the initiatives required to be taken by various Central Ministries/Departments in the wake of natural calamities, sets down the procedure and determines the focal points in the administrative machinery.

12.2.5. State Relief Manuals

Each State Government has relief manuals/codes which identify that role of each officer in the State for managing the natural disasters. These are reviewed and updated periodically based on the experience of managing the disasters and the need of the State.

12.2.6. Funding Mechanisms

The policy and the funding mechanism for provision of relief assistance to those affected by natural calamities is clearly laid down. These are reviewed by the Finance Commission appointed by the Government of India every five years. The Finance Commission makes recommendation regarding the division of tax and non-tax revenues between the Central and the State Governments and also regarding policy for provision of relief assistance and their share of expenditure thereon. A Calamity Relief Fund (CRF) has been set up in each State as per the recommendations of the Eleventh Finance Commission. The size of the Calamity Relief Fund has been fixed by the Finance Commission after taking into account the expenditure on relief and rehabilitation over the past 10 years. The Government of India contributes 75% of the corpus of the Calamity Relief Fund in each State. 25% is contributed to by the State. Relief assistance to those affected by natural calamities is granted from the CRF. Overall norms for relief assistance are laid down by a national committee with representatives of States as members. Different States can have Statespecific norms to be recommended by State level committee under the Chief Secretary. Where the calamity is of such proportion that the funds available in the CRF will not be sufficient for provision of relief, the State seeks assistance from the National Calamity Contingency Fund (NCCF) - a fund created at the Central Government level. When such requests are received, the requirements are assessed by a team from the Central Governemnt and thereafter the assessed requirements are cleared by a High Level Committee chaired by the Deputy Prime Minister. In brief, the institutional arrangements for response and relief are wellestablished and have proved to be robust and effective.

In the federal set up of India, the basic responsibility for undertaking rescue, relief and rehabilitation measures in the event of a disaster is that of the State Governemnt concerned. At the State level, response, relief and rehabilitation are handled by Departments of Relief & Rehabilitation. The State Crisis Management Committee set up under the Chairmanship of Chief Secretary who is the highest executive functionary in the State. All the concerned Departments and organisations of the State and Central Government Departments located in the State are represented in this Committee. This Committee reviews the action taken for response and relief and gives guidelines/directions as necessary. A control room is established under the Relief Commissioner.

The control room is in constant touch with the climate monitoring/forecasting agencies and monitors the action being taken by various agencies in performing their responsibilities. The district level is the key level for disaster management and relief activities. The Collector/ Dy. Commissioner is the chief administrator in the district. He is the focal point in the preparation of district plans and in directing, supervising and monitoring calamities for relief. A District Level Coordination and Relief Committee is constituted and is headed by the Collector as Chairman with participation of all other related government and non governmental agencies and departments in addition to the elected representatives.

The Collector is required to maintain close liaison with the district and the State Governments as well as the nearest units of Armed Forces/Central police organisations and other relevant Central Government organisations like Ministries of Communications, Water Resources, Drinking Water, Surface Transport, who could supplement the efforts of the district administration in the rescue and relief operations. The efforts of the Government and non-governmental organisations for response and relief and coordinated by the Collector/Dy. Commissioner. The District Magistrate/Collector and Coordination Committee under him reviews preparedness measures prior to a impending hazard and coordinate response when the hazard strikes. As all the Departments of the State Government and district level report to the Collector, there is an effective coordination mechanism ensuring holistic response.

12.2.7. New Institutional Mechanisms

As has been made clear above, the existing mechanisms had based on post-disaster relief and rehabilitation and they have proved to be robust and effective mechanisms in addressing these requirements. The changed policy/approach, however, mandates a priority to full disaster aspects of mitigation, prevention and preparedness and new institutional and policy mechanisms are being put in place to address the policy change.

It is proposed to constitute a National Emergency Management Authority at the National level. The High Powered Committee on Disaster Management which was set up in August, 1999 and submitted its Report in October, 2001, had inter alia recommended that a separate Department of Disaster Management be set up in the Government of India. It was, however, felt that conventional Ministries/Departments have the drawback of not being flexible enough specially in terms of the sanction procedures. The organisation at the Apex level will have to be multi-disciplinary with experts covering a large number of branches. The National Emergency Management Authority has, therefore, been proposed as a combined Secretariat/Directorate structure – a structure which will be an integral part of the Government and, therefore, will work with the full authority of the Government while, at the same time, retaining the flexibility of a field organisation. The National Emergency Management Authority will be headed by an officer of the rank of Secretary/Special Secretary to the Government in the Ministry of Home Affairs with Special Secretaries/Additional Secretaries from the Ministries/ Departments of Health, Water Resources, Environment & Forests, Agriculture, Railways, Atomic Energy, Defence, Chemicals, Science & Technology, Telecommunications, Urban Employment and Poverty Alleviation, Rural Development and India Meteorological Department as Members of the Authority. The Authority would meet as often as required and review the status of warning systems, mitigation measures and disaster preparedness. When a disaster strikes, the Authority will coordinate disaster management activities. The

Authority will be responsible for:-

i) Coordinating/mandating Government's policies for disaster reduction/mitigation.

ii) Ensuring adequate preparedness at all levels in order to meet disasters.

iii) Coordinating response to a disaster when it strikes.

iv) Coordination of post disaster relief and rehabilitation.

The National Emergency Management Authority will have a core permanent secretariat with three divisions – one for Disaster Prevention, Mitigation & Rehabilitation, the other for Preparedness and the third for Human Resource Development.

At the State level, as indicated in para disaster management was being handled by the Departments of Relief & Rehabilitation. As the name suggests, the focus was almost entirely on post-calamity relief. The Government of India is working with the State Governments to convert the Departments of Relief &

Rehabilitation into Departments of Disaster Management with an enhanced area of responsibility to include mitigation and preparedness apart from their present responsibilities of relief and rehabilitation. The changeover has already happened in eight State Governments/Union Territory Administrations. The change is under process in other States.

The States have also been asked to set up Disaster Management Authorities under the Chief Minister with Ministers of relevant Departments [Water Resources, Agriculture, Drinking Water Supply, Environment & Forests, Urban Development, Home, Rural Development etc.] as members. The objective of setting up an Authority is to ensure that mitigation and preparedness is seen as the joint responsibility of all the Departments concerned and disaster management concerns are mainstreamed into their programmes. This holistic and multidisciplinary approach is the key to effective mitigation.

At the district level, the District Magistrate who is the chief coordinator will be the focal point for coordinating all activities relating to prevention, mitigation and preparedness apart from his existing responsibilities pertaining to response and relief. The District Coordination and Relief Committee is being reconstituted/ re-designated into Disaster Management Committees with officers from relevant departments being added as members. Because of its enhanced mandate of mitigation and prevention, the district heads and departments engaged in development will now be added to the Committee so that mitigation and prevention is mainstreamed into the district plan. The existing system of drawing up preparedness and response plans will continue. There will, however, also be a long term mitigation plan. District Disaster Management Committees have already been constituted in several districts and are in the process of being constituted in the remaining multi-hazard prone districts.

Similarly, we are in the process of creating Block/ Taluq Disaster Management Committees in these 169 multi-hazard prone districts in 17 States. At the village level, in 169 multi-hazard prone districts, we are constituting Disaster Management Committees and Disaster Management Teams. Each village will have a Disaster Management Plan. The process of drafting the plan has already begun. The Disaster Management Committee which draws up the plans consists of elected representatives at the village level, local authorities, Government functionaries including doctors/paramedics of primary health centres located in the village, primary school teachers etc. The plan encompasses prevention, mitigation and preparedness measures. The Disaster Management Teams at the village level will consist of members of voluntary organisations like Nehru Yuvak Kendra and other non-governmental organisations as well as able bodied volunteers from the village. The teams are provided basic training in evacuation, search and rescue etc. The Disaster Management Committee will review the disaster management plan at least once in a year. It would also generate awareness among the people in the village about dos' and don'ts for specific hazards depending on the vulnerability of the village. A large number of village level Disaster Management Committees and Disaster Management Teams have already been constituted.

The States have been advised to enact Disaster Management Acts. These Acts provide for adequate powers for authorities coordinating mitigation, preparedness and response as well as for mitigation/ prevention measures required to be undertaken. Two States [Gujarat & Madhya Pradesh] have already enacted such a law. Other States are in the process. The State Governments have also been advised to convert their Relief Codes into Disaster Management Codes by including aspects of prevention, mitigation and preparedness.

In order to further institutionalise the new approach, the Government of India have decided to enunciate a National Policy on Disaster Management. A draft policy has accordingly been formulated and is expected to be put in place shortly. The policy shall inform all spheres of Central Government activity and shall take precedence over all existing sectoral policies. The broad objectives of the policy are to minimise the loss of lives and social, private and community assets because of natural or manmade disasters and contribute to sustainable development and better standards of living for all, more specifically for the poor and vulnerable sections by ensuring that the development gains are not lost through natural calamities/disasters.

The policy notes that State Governments are primarily responsible for disaster management including prevention and mitigation, while the Government of India provides assistance where necessary as per the norms laid down from time to time and proposes that this overall framework may continue. However, since response to a disaster requires coordination of resources available across all the Departments of the Government, the policy mandates that the Central Government will, in conjunction with the State Governments, seek to ensure that such a coordination mechanism is laid down through an appropriate chain of command so that mobilisation of resources is facilitated.

The broad features of the draft national policy on disaster management are enunciated below:-

i) A holistic and pro-active approach for prevention, mitigation and preparedness will be adopted for disaster management.

ii) Each Ministry/Department of the Central/State Government will set apart an appropriate quantum of funds under the Plan for specific schemes/projects addressing vulnerability reduction and preparedness.

iii) Where there is a shelf of projects, projects addressing mitigation will be given priority. Mitigation measures shall be built into the on-going schemes/programmes

iv) Each project in a hazard prone area will have mitigation as an essential term of reference. The project report will include a statement as to how the project addresses vulnerability reduction.

v) Community involvement and awareness generation, particularly that of the vulnerable segments of population and women has been emphasized as necessary for sustainable disaster risk reduction. This is a critical component of the policy since communities are the first responders to disasters and, therefore, unless they are empowered and made capable of managing disasters, any amount of external support cannot lead to optimal results.

vi) There will be close interaction with the corporate sector, nongovernmental organisations and the media in the national efforts for disaster prevention/ vulnerability reduction.

vii) Institutional structures/appropriate chain of command will be built up and appropriate training imparted to disaster managers at various levels to ensure coordinated and quick response at all levels; and development of inter-State arrangements for sharing of resources during emergencies.

viii) A culture of planning and preparedness is to be inculcated at all levels for capacity building measures.

ix) Standard operating procedures and disaster management plans at state and district levels as well as by relevant central government departments for handling specific disasters will be laid down.

x) Construction designs must correspond to the requirements as laid down in relevant Indian Standards.

xi) All lifeline buildings in seismic zones III, IV & V – hospitals, railway stations, airports/airport control towers, fire station buildings, bus stands major administrative centres will need to be evaluated and, if necessary, retro-fitted.

xii) The existing relief codes in the States will be revised to develop them into disaster management codes/ manuals for institutionalizing the planning process with particular attention to mitigation and preparedness.

With the above mentioned institutional mechanism and policy framework in position and the actions taken to implement the policy guidelines, it is expected that the task of moving towards vulnerability reduction will be greatly facilitated.

12.3. Early Warning System

12.3.1. Cyclone Forecasting

Tropical Cyclones are intense low pressure systems which develop over warm sea. They are capable of causing immense damage due to strong winds, heavy rains and storm surges. The frequency of the TC in the Bay of Bengal is 4 to 5 times more than in the Arabian Sea. About 35% of initial disturbances in the north Indian ocean reach TC stage of which 45% become severe.

Indian Meteorological Department (IMD) is mandated to monitor and give warnings regarding Tropical Cyclone (TC). Monitoring process has been revolutionised by the advent of remote sensing techniques. A TC intensity analysis and forecast scheme has been worked out using satellite image interpretation techniques which facilitate forecasting of storm surges.

Data resources are crucial to early forecasting of cyclones. Satellite based observations are being extensively utilised. Satellite integrated automated weather stations have been installed on islands, oilrigs and exposed coastal sites. Buoys for supplementing the surface data network in the tropical ocean have been deployed. The Government have also started a National Data Buoy Programme. A set of 12 moored buoys have been deployed in the northern Indian Ocean to provide meteorological and oceanographic data.

Dynamic forecasting of TCs requires knowledge of the vertical structure of both the Cyclone and the surrounding environment. The rawin sonde remains the principal equipment for sounding. The Doppler Radar wind profiler provides hourly soundings. A mesosphere, stratosphere, troposphere (MST) radar has also been installed at Thirupatti. Another profiler is being developed and will be deployed at IMD Pune. Another important source of upper level data is the aircraft reports. Increasing number of commercial jet aircraft are equipped with the Aircraft Meteorological Data Relay system. This data is being made available is also being used by the IMD for analysis and predictions.

Radars have been used to observe TCs since long. Surveillance of the spiral rain bands and the eye of the TC is an important function of the coastal radars. 10 Cyclones Detection Radars have already been installed. These radars are providing useful estimates of storm centres upto a range 300-400 Km. Doppler radars provide direct measurements of wind fields in TCs. Due to range limitation, Doppler wind estimates are usually within a range of about 100 Km. IMD has deployed Doppler radars at 3 sites on the east coast. Another set of 3

Doppler radars are being deployed in Andhra Pradesh in near future.

The meteorological satellite has made a tremendous impact on the analysis of cyclones. All developing cloud clusters are routinely observed through satellite cloud imagery & those showing signs of organisation are closely monitored for signs of intensification. TC forecasters everywhere use the Dvorak technique to estimate storm location and intensity. It has been found to provide realistic estimates for TCs in the Bay of Bengal as well as Arabian Sea. INSAT data has also been used to study the structures of different TCs in the Bay of Bengal. IMD is also producing Cloud Motion Vectors (CMVs). Very High Resolution Radiometer (VHRR) payload onboard INSAT –2E which have been improved upon to provide water vapor channel data in addition to VIS & IR onboard INSAT – 2E. A separate payload known as Charged Couple Device (CCD) has also been deployed onboard this satellite.

The goal of any warning system is to maximise the number of people who take appropriate and timely action for the safety of life and property. All warning systems start with detection of the event and with people getting out of harm's way. Such warning systems encompass three equally important elements namely; Detection and Warning; Communication; and Response.

The two stage warning system has been in existence since long in IMD. Recently it has been improved upon by introducing two more stages - the 'Pre-Cyclone watch' and the 'post-landfall Scenario'. This four stage warning system meets the requirements of Public Administrators and Crisis Managers. The 'Pre-Cyclone Watch' stage, contains early warning about the development of a cyclonic disturbance in the form of monsoon depression which has a potential to threaten the coast with cyclone force winds. The coastal stretch likely to be affected is identified. This early warning bulletin is issued by the IMD before the Cyclone-Alert Stage. This provides enough lead time for the crisis managers to undertake preparedness actions.

After the early warning on the 'Pre-Cyclone Watch' the Collectors of coastal and few immediate interior districts and the Chief Secretary of the concerned maritime State are warned in two stages, whenever any coastal belt is expected to experience adverse weather (heavy rain/gales/tidal wave) in association with a cyclonic storm or a depression likely to intensify into a cyclonic storm.

The second stage of "Cyclone Alert" is sounded 48 hours in advance of the expected commencement of adverse weather over the coastal areas. Forecasts of commencement of strong winds, heavy precipitation along the coast in association with arrival of cyclone are issued at the alert stage. Landfall point is usually not identified at this stage. The third stage warning known as "Cyclone Warning" is issued 24 hours in advance. Landfall point is forecast in this stage of cyclone warning. In addition to the forecasts for heavy rains and strong winds, the storm surge forecast is also issued. Since the storm surge is the biggest killer so far as the devastating attributes of a storm are concerned, information in this regard is most critical for taking follow up action for evacuation from the low lying areas likely to be affected by the storm.

After the landfall of the cyclone the strong winds with gale force speeds continue over certain interior districts of the maritime States hit by the cyclone. To take cognisance of that, a fourth stage known as 'Post-landfall Scenario Stage' is now identified usually as a part of the 'Cyclone Warning Stage' either at the time of landfall of the disturbance or about twelve hour in advance of it. It includes warnings of strong winds and heavy rains likely to be encountered in the interior districts.

For communications, the IMD makes use of 97 point-to-point teleprinter links connecting different field offices. Switching computers have been provided at 5 Regional Centres. These computers are linked to the central Regional Telecom Hub Computer at New Delhi. In addition, 69 centres have been provided with 85 telex connections. Besides, 27 field offices have been provided with Radio Teletype facility. IMD also utilises VSAT technology which has been installed at field offices. In addition, there are a number of HF/RT and VHF links.

Cyclone warnings are communicated to Crisis Managers and other concerned organisations by high priority telegrams, telex, telephones and Police wireless. Cyclone warning are provided by the IMD from the Area Cyclone Warning Centres (ACWCs) at Calcutta, Chennai and Mumbai and Cyclone Warning Centers (CWCs) at Vishakhapatnam, Bhubaneswar and Ahmedabad. There is also a Satellite based communication system called the Cyclone Warning Dissemination Systems (CWDS) for transmission of warnings. There are 250 such cyclone-warning sets installed in the cylone prone areas of east and west coast. The general public, the coastal residents and fishermen, are also warned through the Government mechinery and broadcast of warnings through AIR and Television.

12.3.2. Flood Forecasting

Flooding is caused by the inadequate capacity within the banks of the rivers to contain the high flow brought down from the upper catchments due to heavy rainfall. It is also caused by accumulation of water resulting from heavy spells of rainfall over areas, which have got poor drainage characteristics.

Flooding is accentuated by erosion and silting leading to meandering of the rivers in plains and reduction in carrying capacity of the river channel. It is also aggravated by earthquakes and land slides, leading to changes in river course and obstructions to flow. Synchronisation of floods in the main rivers and tributaries and retardation of flow due to tidal effects lead to major floods. Cyclones bring in their wake considerable loss of life and property.

The flood forecasting and warning system is used for alerting the likely damage centers well in advance of the actual arrival of floods, to enable the people to move and also to remove the moveable property to safer places or to raised platforms specially constructed for the purpose.

A beginning in scientific flood forecasting was made in November, 1958 by Central Water Commission (then known as Central Water & Power Commission) when a Flood Forecasting Centre was set up at its Headquarters, at Delhi, for giving timely Forecasts and Warnings of the incoming floods to the villages located in the river areas around the National Capital, Delhi. The network has been expanding and by now the Flood Forecasting Network of the Central Water Commission(CWC) covers all the major flood prone inter-State river basins in the country.

At present there are 166 flood forecasting stations on various rivers in the country which includes 134 level forecasting and 32 inflow forecasting stations, river-wise break up of which is as under:

Sl.No.	*Name of River Systems*	*No. of Flood Forecasting Stations*		
		Level	*Inflow*	*Total*
1.	Ganga & Tributaries	71	14	85
2.	Brahmaputra & Tributaries	27	-	27
3.	Barak-System	2	-	2
4.	Eastern-Rivers	8	1	9

5. Mahanadi	2	1	3
6. Godavari	13	4	17
7. Krishna	2	6	8
8. West Flow Rivers	9	6	15
Total	134	32	166

The Flood Forecasting Network covers the 14 States and one Union Territory in addition to NCT of Delhi. State-wise number of flood forecasting centres are as under :

Sl.No.	*Name of River Systems*	*No. of Flood Forecasting Stations*		
		Level	*Inflow*	*Total*
1.	Andhra Pradesh	8	7	15
2.	Assam	23	-	23
3.	Bihar	32	-	32
4.	Chhattishgarh	01	-	01
5.	Gujarat	6	4	10
6.	Haryana	-	01	01
7.	Jharkhand	-	04	04
8.	Karnataka	01	03	04
9.	Madhya Pradesh	2	-	02
10.	Maharashtra	5	02	07
11.	Orissa	10	01	11
12.	Uttaranchal	01	02	03
13.	Uttar Pradesh	31	04	35
14.	West Bengal	11	03	14
15.	Dadra & Nagar Haveli	01	01	02
16.	N.C.T. of Delhi	02	-	02
	All India total	134	32	166

The Flood Forecasting involves the following four main activities :-

(i) Observation and collection of hydrological and hydro-meteorological data;

(ii) Transmission of Data to Forecasting Centres;

(iii) Analysis of data and formulation of forecast; and

(iv) Dissemination of forecast.

On an average, 6000 forecasts at various places in the country are issued during the monsoon season every year. The analysis of the forecasts issued during the last 25 years (1978 to 2002) indicates that accuracy of forecasts has consistently increased from around 81% to 98%. Forecast is considered accurate if forecast water level is within ± 15 cm. of actual water level of the inflow forecast (i.e. discharge) is with in ± 20% of actual discharge.

In monitoring the floods, severity of floods are placed in the following four categories by the central Water Commissions.

(i) *Low flood stage:* It is that flood situation when the water level of the river is flowing between warning level and danger level of the forecasting stations.

(ii) *Medium flood stage:* The river is called in medium floods when its water level is at or above the danger level of the forecasting station but below 0.50 of its highest flood level (HFL).

(iii) *High flood stage:* When the water level of the river is below the HFL but within 0.50 m. of the HFL of the forecasting stations.

(iv) *Unprecedented flood stage :* The river is called in unprecedented floods when it attains water level equal to or above its previous HFL at any forecasting station.

A computerised monitoring system has been developed under which daily water levels as observed at 0800 hrs. and forecasts issued by field units are transmitted to CWC headquarters in New Delhi. Based on the compilation of all such data received from field divisions, daily water level and flood forecast bulletins in two parts for stage and for inflow forecasting stations respectively.

Special Yellow Bulletins are issued whenever the river stage at the forecasting site attains a level within 0.50 m of its previous HFL. Red Bulletins highlighting security of the problem are also issued whenever the water level at the forecasting stations equals or exceeds previous HFL.

12.4. Disaster Prevention and Mitigation

The Yokohama message emanating from the international decade for natural disaster reduction in May, 1994 underlined the need for an emphatic shift in the strategy for disaster mitigation. It was inter alia stressed that disaster prevention, mitigation, preparedness and relief are four elements which contribute to and gain, from the implementation of the sustainable development policies. These elements alongwith environmental protection and sustainable development, are closely inter related. Therefore, nations should incorporate them in their development plans and ensure efficient follow up measures at the community, sub-regional, regional, national and international levels. The Yokohama Strategy also emphasised that disaster prevention, mitigation and preparedness are better than disaster response in achieving the goals and objectives of vulnerability reduction. Disaster response alone is not sufficient as it yields only temporary results at a very high cost. Prevention and mitigation contribute to lasting improvement in safety and are essential to integrated disaster management.

The Government of India have adopted mitigation and prevention as essential components of their development strategy. The Tenth Five Year Plan document has a detailed chapter on Disaster Management. The plan emphasizes the fact that development cannot be sustainable without mitigation being built into developmental process. Each State is supposed to prepare a plan scheme for disaster mitigation in accordance with the approach outlined in the plan. In brief, mitigation is being institutionalised into developmental planning.

As indicated in the earlier chapter, the Finance Commission makes recommendations with regard to devolution of funds between the Central Government and State Governments as also outlays for relief and rehabilitation. The earlier Finance Commissions were mandated to look at relief and rehabilitation. The Terms of Reference of the Twelfth Finance Commission have been changed and the Finance Commission has been mandated to look at the requirements for mitigation and prevention apart from its existing mandate of looking at relief and rehabilitation. A Memorandum has been submitted to the Twelfth Finance Commission after consultation with States. The Memorandum proposes a Mitigation Fund.

The Government of India have issued guidelines that where there is a shelf of projects, projects addressing mitigation will be given a priority. It has also been mandated that each project in a hazard prone area will have disaster prevention/mitigation as a term of reference and the project document has to reflect as to how the project addresses that term of reference.

Measures for flood mitigation were taken from 1950 onwards. As against the total of 40 million hectares prone to floods, area of about 15 million hectares have been protected by construction of embankments. A number of dams and barrages have been constructed. The State Governments have been assisted to take up mitigation programmes like construction of raised platforms etc. Floods continue to be a menace however mainly because of the huge quantum of silt being carried by the rivers emanating from the Himalayas. This silt has raised the bed level in many rivers to above the level of the countryside. Embankments have also gives rise to problems of drainage with heavy rainfall leading to water logging in areas outside the embankment.

Due to erratic behaviour of monsoons, both low and medium rain fall regions, which constitute about 68% of the total area, are vulnerable to periodical droughts. Our experience has been that almost every third year is a drought year. However, in some of the States, there may be successive drought years enhancing the vulnerability of the population in these areas. Local communities have devised indigenous safety mechanisms and drought oriented farming methods in many parts of the country. From the experience of managing the past droughts particularly the severe drought of 1987, a number of programmes have been launched by the Government to mitigate the impact of drought in the long run. These programmes include Drought Prone Area Programme (DPAP), Desert Development Programme (DDP), National Watershed Development Project for Rainfed Areas (NWDPRA), Watershed Development Programme for Shifting Cultivation (WDPSC), Integrated Water Development Project (IWDP), Integrated Afforestation and Eco-development Project Scheme (IAEPS).

A comprehensive programme has been taken up for earthquake mitigation. Although, the BIS has laid down the standards for construction in the seismic zones, these were not being followed. The building construction in urban and suburban areas is regulated by the Town and Country Planning Acts and Building Regulations. In many cases, the Building regulations do not incorporate the BIS codes. Even where they do, the lack of knowledge regarding seismically safe construction among the architects and engineers as well as lack of awareness regarding their vulnerability among the population led to most of the construction in the urban/sub-urban areas being without reference to BIS standards. In the rural areas, the bulk of the housing is non-engineered construction. The mode of construction in the rural areas has also changed from mud and thatch to brick and concrete construction thereby increasing the vulnerability. The increasing population has led to settlements in vulnerable areas close to the river bed areas which are prone to liquefaction. The Government have moved to address these issues.

A National Core Group for Earthquake Mitigation has been constituted consisting of experts in earthquake engineering and administrators. The Core Group has been assigned with the responsibility of drawing up a strategy and plan of action for mitigating the impact of earthquakes; providing advice and guidance to the States on various aspects of earthquake mitigation; developing/ organising the preparation of handbooks/pamphlets/ type designs for earthquake resistant construction; working out systems for assisting the States in the seismically vulnerable zones to adopt/integrate appropriate Bureau of Indian Standards codes in their building byelaws; evolving systems for training of municipal engineers as also practicing architects and engineers in the private sector in the salient features of Bureau of Indian Standards codes and the amended byelaws; evolving a system of certification of architects/ engineers for testing their knowledge of earthquake resistant construction; evolving systems for training of masons and carry out intensive awareness generation campaigns.

A Committee of experts has been constituted to review the building byelaws. The State Governments have been advised to ensure rigorous enforcement of existing bye laws. A national programme for capacity building for earthquake mitigation has been finalised for imparting training to 10000 engineers in public and private sectors. Since earthquake engineering is not a part of course curriculum in engineering colleges at undergraduate level at present , it is proposed to select 3 to 4 leading engineering colleges in each State and train the faculty members of the civil engineering departments in earthquake engineering at the Indian Institutes of Technology and few other apex level institutes which have the requisite capabilities.

These faculty members will take up training of municipal engineers as well as the training of engineers/ architects in the private sector in RCC and masonry construction. The first phase of this programme for imparting training to 10000 engineers will be completed within a period of three years. The trained faculty members of the leading engineering colleges will also assist the State Governments in the detailed evaluation of lifeline buildings and their retrofitting, wherever necessary.

It has been decided to include earthquake engineering education in the engineering colleges at undergraduate level. The course curriculum for this purpose has already been finalised by a group of experts taken from IITs and will be introduced in the engineering colleges within an year. A system of special audit of buildings is being put in place with a view to ensuring that the new constructions conform to the latest building byelaws, which have been reviewed and revised recently by Bureau of Indian Standards.

While these mitigation measures will take care of the new constructions, the problem of unsafe existing buildings stock would still remain. It will not be possible to address the entire existing building stock, therefore the life line buildings like hospitals, schools or buildings where people congregate like cinema halls, multi-storied apartments are being focussed on. The States have been advised to have these buildings assessed and where necessary retrofitted. The Ministry of Finance have been requested to advise the financial institutions to give loans for retrofitting on easy terms. Insofar as the private housing stock is concerned emphasis is placed on awareness generation.

An earthquake mitigation project has been finalised for reducing the vulnerability to earthquakes. The programme includes detailed evaluation and retrofitting of lifeline buildings such as hospitals, schools, water and power supply units, telecommunication buildings, airports/airport control towers, railway stations, bus stands and important administrative buildings. The programme also includes training of more than one hundred thousand masons for earthquake resistant constructions. Besides, assistance will be provided under this project to the State Governments to put in place appropriate techno legal regime.

An accelerated urban earthquake vulnerability reduction programme has been taken up in 38 cities in seismic zones III, IV & V with population of half a million and above. Sensitization workshop for engineers/ architects, government functionaries and voluntary organisations have already been held in 36 of the 38 cities. Disaster mitigation and preparedness plans are under preparation in these cities. Awareness generation campaign has already been undertaken. The orientation courses for engineers and architects have been organised to impart knowledge about seismically safe construction and implementation of BIS norms. This programme will be further extended to 166 earthquake prone districts in seismic zones IV & V.

Rural housing and community assets for vulnerable sections of the population are created at a fairly large scale by the Ministry of Rural Development under the Indira Awas Yojna(IAY) and Sampooran Grameen Rojgar

Yojna (SGRY). About 250 thousand small but compact units are constructed every year, besides community assets such as community centres, recreation centres, anganwadi centres etc. Technology support is provided by about two hundred rural housing centres spread over the entire country. The Ministry of Rural Development are now under the process of revising their guidelines for construction of such dwelling units by incorporating appropriate earthquake/cyclone resistant features. Training to the functionaries in the rural housing centres will be organised through the Ministry of Home Affairs . This initiative is expected to go a long way for the construction and popularisation of seismically safe construction at village/block level.

A National Core Group on Cyclone Monitoring & Mitigation has been constituted. Experts from Indian Meteorological Department, National Centre for Medium Range Weather Forecasting, Central Water Commission, National Remote Sensing Agency and Indian Space Research Organisation have been made the Members of the Core Group, besides administrators from the relevant Ministries/Departments and State Governments vulnerable to cyclones.

The Group has been assigned with the responsibility of looking warning protocols for cyclones; coordination mechanism between different Central and State Ministries/Departments/Organisations; mechanism for dissemination of warning to the local people and; cyclone mitigation measures required to be taken for the coastal States. The Group will also suggest short-term and long-term measures on technology upgradation.

A cyclone mitigation project has been formulated. The project inter alia includes components on strengthening of monitoring/warning systems, coastal shelter belt plantation, mangrove plantation, construction of cyclone shelters, storm surge modeling and water envelope studies. The focus will be on regeneration of coastal shelter belt plantation and mangrove plantation where these have degenerated. The location of the cyclone shelters will be decided in such a manner that no person in the vulnerable zone is required to walk more than two kilometers to reach a cyclone shelter. The cyclone shelters will be multi purpose units to be run as schools or community centres in normal times and will have capacity to house 3000 to 5000 persons with adequate number of toilets, community kitchen and other facilities. Areas will be identified for providing shelter to livestock.

In the engineering designs for construction, special attention will be paid to the attachment of roof to the dwelling units so as to make such units cyclone proof, besides incorporating earthquake resistant features. The project will be taken up shortly and is expected to be completed over a period of five years.

A Disaster Risk Management Programme has been taken up with the assistance from UNDP, USAID and European Union in 169 most hazard prone districts in 17 States including all the 8 North Eastern State. The implementation of the project commenced from October, 2002 and is expected to be concluded by December, 2007. The programme components include awareness generation and public education, preparedness, planning and capacity building, developing appropriate policies, institutional, administrative, legal and techno-legal regime at State, District, Block, village, urban local body and ward levels for vulnerability reduction.

Under this programme Disaster Management Plans have been prepared for about 3500 villages, 250 Gram Panchayat, 60 blocks and 15 districts. Elected representatives of over 8000 Panchayati Raj Institutions have already been trained, besides imparting training to Members of voluntary organisations. Over 20000 Government functionaries have been trained in disaster mitigation and preparedness at different levels. About 600 engineers and 220 architects have been trained under this programme in vulnerability assessment of lifeline buildings. Training is being imparted to master trainers under the programme. More than 600 master trainers and 1000 teachers have already been trained in different districts in disaster mitigation.

Disaster Management Committees consisting of elected representatives, civil society members, Civil Defence volunteers and Government functionaries have been constituted at all levels including village/urban local body/ward levels. Disaster Management Teams have been constituted in villages and are being imparted training in basic functions of first aid, rescue, evacuation and related issues. The thrust of the programme is to build up capabilities of the community since the community is invariably the first responder.

During the last 15 months, it has been experienced that the capacity building of the community has been very helpful even in normal situations when isolated instances of drowning, burns etc. take place. With the creation of awareness generation on disaster mitigation, the community will be able to function as a well-knit unit in case of any emergency. Mock drills are carried out from time to time under the close supervision of Disaster Management Committees.

The Disaster Management Committees and Disaster Management Teams have been established by notifications issued by the State Governments which will ensure that the entire system is institutionalised and does not disintegrate after the conclusion of the programme. The key points being stressed under this programme are the need to ensure sustainability of the programme, development of training modules; manuals and codes, up-scaling partnerships in excellence, focused attention to awareness generation campaigns; institutionalisation of disaster management committees and disaster management teams, disaster management plans and mock-drills and establishment of techno-legal regimes.

Human Resource Development at all levels is critical to institutionalisation of disaster mitigation strategy. The National Centre for Disaster Management at the national level has been upgraded and designated as the National Institute of Disaster Management. It is being developed as a Regional Centre of Excellence in Asia.

The National Institute of Disaster Management will develop training modules at different levels, undertake training of trainers and organise training programmes for planners, administrators and command functionaries. Besides, the other functions assigned to the National Institute of Disaster Management include development of exhaustive National level information base on disaster management policies, prevention mechanisms, mitigation measures; formulation of disaster management code and providing consultancy to various States in strengthening their disaster management systems and capacities as well as preparation of disaster management plans and strategies for hazard mitigation and disaster response.

Disaster Management faculties have already been created in 29 State level training institutes located in 28 States. These faculties are being directly supported by the Ministry of Home Affairs. The State Training Institutions take up several focused training programmes for different target groups within the State.

The Disaster Management faculties in these Institutes are being further strengthened so as to enable them to develop as Institutes of Excellence for a specific disaster. This system has already been institutionalised and is being further strengthened so as to make it a focal point in each State for development of human resources in disaster mitigation and preparedness. Assistance to the

State level training institutes will be provided by the National Institute of Disaster Management in the development of training/capsules training modules for different functionaries at different levels.

Large-scale awareness generation bringing out specific do's and don'ts is crucial to disaster mitigation. A Steering Committee on mass-media campaign has been constituted for this purpose. The Committee is in the process of developing a profile for taking up mass media campaign through audio, video and print media as well as publicity through pamphlets, posters, bus back panels at all levels. The posters would be prominently displayed at buildings like Primary Health Centres, Community Centres, schools and such other places where villagers normally congregate for community activity. The Corporate sector is also being associated with the dissemination of campaign.

Disaster management as a subject in Social Sciences has been introduced in the school curriculum for Class VIII from the current academic year. The Central Board of Secondary Education which has introduced the curriculum runs a very large number of schools throughout the country and the course curriculum is invariably followed by the State Boards of Secondary Education. Several State Governments have already introduced the same curriculum in Class VIII from the current academic year. Syllabus for Class IX and X has been finalised and will be introduced in the course curriculum from April, 2004 and April, 2005 respectively.

In order to assist the State Governments in capacity building and awareness generation activities and to learn from past experiences including sharing of best practices, the Ministry of Home Affairs has compiled/prepared a set of resource materials developed by various organisations/institutions to be replicated and disseminated by State Governments based on their vulnerabilities after translating it into the local languages. The voluminous material which runs in about 10000 pages has been divided into 4 broad sections in 7 volumes. These sections coverplanning to cope with disasters; education and training; construction toolkit; and information, education and communication toolkit including multi-media resources on disaster mitigation and preparedness. The Planning section contains material for analysing a community's risk, development of Preparedness. mitigation and disaster management plans, coordinating available resources and implementing measures for risk reduction.

The model bye-laws, DM Policy, Act and model health sector plan have also been included. Education and Training includes material for capacity building and upgradation of skills of policy makers, administrators, trainers, engineers etc. in planning for and mitigating against natural disasters. Basic and detailed training modules in disaster preparedness have been incorporated along with training methodologies for trainers, for community preparedness and manuals for training at district, block, panchayat and village levels.

For creating a disaster-resistant building environment, the Construction Toolkit addresses the issue of seismic resistant construction and retrofitting of existing buildings. BIS Codes, manuals and guidelines for RCC, Masonry and other construction methodologies as also for repair and retrofitting of masonry and low-rise buildings have been included.

IEC material seeks to generate awareness to induce mitigation and preparedness measures for risk reduction Material and strategies used by various States and international organisations, including tips on different hazards, have been incorporated along with multi-media CDs on disasters. The material has been disseminated to all the State Governments/UT Administrations with the request to have the relevant material, based on the vulnerability of each district, culled out, translated into local languages and disseminate it widely down to the village level.

The various prevention and mitigation measures outlined above are aimed at building up the capabilities of the communities, voluntary organisations and Government functionaries at all levels. Particular stress is being laid on ensuring that these measures are institutionalised considering the vast population and the geographical area of the country.

This is a major task being undertaken by the Government to put in place mitigation measures for vulnerability reduction. This is just a beginning. The ultimate goal is to make prevention and mitigation a part of normal day-to-day life. The above mentioned initiatives will be put in place and information disseminated over a period of five to eight years. We have a firm conviction that with these measures in place, we could say with confidence that disasters like Orissa cyclone and Bhuj earthquake will not be allowed to recur in this country; at least not at the cost, which the country has paid in these two disasters in terms of human lives, livestock, loss of property and means of livelihood.

12.5. Preparedness

Mitigation and preparedness measures go hand in hand for vulnerability reduction and rapid professional response to disasters. The Bhuj earthquake in January, 2001 brought out several inadequacies in the system. The search and rescue teams had not been trained professionally; specialised dog squad to look for live bodies under the debris were not available; and there was no centralised resource inventory for emergency response. Although army played a pivotal role in search and rescue and also set up their hospital after the collapse of Government hospital at Bhuj, the need for fully equipped mobile hospitals with trained personnel was felt acutely. Despite these constraints, the response was fairly well organised. However, had these constraints been taken care of before hand, the response would have been even more professional and rapid which may have reduced the loss of lives.

Specialist search and rescue teams from other countries did reach Bhuj. However, precious time was lost and even with these specialist teams it was not possible to cover all severely affected areas as quickly as the Government would have desired. It was, therefore, decided that we should remove these inadequacies and be in a stage of preparedness at all times.

The Central Government are now in the process of training and equipping 96 specialist search and rescue teams, with each team consisting of 45 personnel including doctors, paramedics, structural engineers etc. Ten teams have already been trained. These teams will be located at various centres around the country for specialised response. These teams will have the latest equipment as also dog squads for locating survivors in the debris.

Apart from specialist search & rescue units, it has been decided that personnel of Central Police Organisations should also be imparted training in search and rescue so that they can be requisitioned to the site of incident without loss of time. Pending arrival of the specialist teams, the battalions located near the site of incident would be deployed immediately. For this purpose, a curriculum has been drawn up and integrated into the training curriculum of CPMFs.

The States have also been advised to set up their own specialist teams for responding to disasters. Assistance will be provided to the State Governments for training their trainers at the national institutes already designated for this purpose. The State Governments' search and rescue teams to be constituted from the State Police will be equipped to meet the requirement. For this purpose, the State Governments have been authorised to utilise 10% of the annual allocation made under the Contingency Relief Fund for purchase of equipments.

Fourteen Regional Response Centres are being set up in different parts of the country . These centres will have response teams and equipment and resources for being able to response to any hazard/calamity in the neighbouring States.

A Steering Committee has been constituted in the Ministry to oversee the creation of capabilities for emergency response.

A 200 bedded mobile hospital, fully trained and equipped is being set up by the Ministry of Health and attached to a leading Government hospital in Delhi. Three additional mobile hospitals with all medical and emergency equipments are proposed to be located in different parts of the country. These mobile hospitals will also be attached to the leading Government hospitals in the country. This will enable the mobile hospitals to extend assistance to the hospitals with which they are attached in normal time. They will be airlifted during emergencies with additional doctors/paramedics taken from the hospitals with which the mobile hospitals are attached to the site of disaster.

It is proposed to purchase dedicated aircraft and helicopters with a view to reducing the response time. The issue is pending for consideration and approval of Empowered Group of Ministers on Disaster Management. Once the airlift facilities are developed for exclusive use for disaster management, it will be possible to provide airlift facilities to specialist search and rescue teams, mobile hospitals and equipments.

In order to professionalise the response, it is proposed to introduce the Incident Command System in the country. This system provides for specialist incident command teams with an Incident Commander and officers trained in different aspects of incident management – logistics, operations, planning, safety, media management etc. The incident Command System has been finalised keeping in view the systems and procedures prevalent in our country by dovetailing it in the existing governmental machinery already in position. The training of trainers in the Incident Command System has already commenced at Lal Bahadur Shastri National Academy of Administration at Mussoori which has been designated as the nodal training institutes for this purpose.

A web-enabled centralised data base for the India Disaster Resource Network has been operationalsed. The network will ensure quick access to resources to minimise response time in emergencies. The list of resources to be updated in the system has been finalised. It has 226 items. About 60,000 records in 481 districts throughout the country have already been uploaded since Ist September, 2003 when the India Disaster Resource Network was formally inaugurated. The system will give, at the touch of the button, location of specific equipments/specialist resources as well as the Controlling authority for that resource so that it can be mobilised for response in the shortest possible time. The data base will be available simultaneously at the district, state and national levels.

The States are being persuaded to set up control rooms/emergency operations centres at the state and district level. Assistance for construction and purchase of equipments for control rooms is being provided. The control rooms, which will function round the clock, will be composite control rooms to look after law and order issues as well as disaster management. Equipments are also being provided for these control rooms under the disaster risk management programme.

Communication is a major bottleneck in case of any major disaster particularly when the traditional network system already in force brake down. In order to strengthen communications, it has been decided that police network (POLNET) will also be used for disaster management.

For this purpose POLNET communication facility will be extended to District Magistrates, Sub Divisional Magistrates as well as the Control Rooms. For emergency communication, mobile satellite based units which can be transported to the site of the disaster are being procured. A group was constituted to draw a comprehensive communication plan for disaster management and the report has since been received. This provides for a dedicated communication system for disaster management with built in redundancies.

The Geographical Information System (GIS) data base is an effective tool for emergency responders to access information in terms of crucial parameters for the disaster affected areas. The crucial parameters include location of the public facilities, communication links and transportation network at national, state and district levels.

The GIS data base already available with different agencies of the Government is being upgraded and the gaps are proposed to be bridged. A project for this purpose is being drawn up with a view to institutionalising the arrangements. The data base will provide multi layered maps on district wise basis. Three maps taken in conjunction with the satellite images available for a particular area will enable the district administration as well as State Governments to carry out hazard zonation and vulnerability assessment, as well as coordinate response after a disaster.

In order to further strengthen the capacity for response, the fire services are proposed to be developed into multi hazard response units as is the normal practice in several other countries. It is proposed to provide rescue tenders in addition to fire tenders to each fire unit and fill up all gaps upto sub-divisional level. Hazmat vans will be provided to State capitals and metropolitan cities. This will necessitate recruitment of additional fire men and drivers and intensive training required to be provided to enable them to function as efficient of all purpose response units. A project for development of fire service into all hazard response units has also been finalised and submitted for obtaining necessary financial approval.

India has a large network of Civil Defence and Home Guards volunteers. The existing strength is about 1.2 million. However, this organisation has not so far been associated with disaster mitigation, preparedness and response functions. It is proposed to revamp the Civil Defence organisation to enable them to discharge a key responsibility in all the facets of disaster management including preparedness. A proposal in this regard has been finalised and is under consideration of the Government.

Standard Operating Procedures are being laid down to ensure all the steps required to be taken for disaster management are put in place. The Standard Operating Procedure will also in encompass response, besides preparedness.

With the development of disaster management committees and disaster management teams at all levels including village/urban local body/ward level, the stage will be set for comprehensive preparedness measures to be taken with active participation of the community and non-governmental organisations.

With the mitigation and preparedness measures outlined in this and earlier section in position, it is expected that natural hazards could be handled more efficiently so as to ensure that these hazards did not get converted into disasters.

Bibliography

Abramovitz, J., "Unnatural Disasters", *World Watch*, Worldwatch Institute, Washington, D.C., 1999

_____________ ., "Averting Natural Disasters" in: Brown, L., Flavin, C., and French, Hillary (eds.), *State of the World 2001*, W.W. Norton & Company, Worldwatch Institute 2001,

Affeltranger, B., "Public Participation in the design of local strategies for flood mitigation and control", *IHP, Technical Documents in Hydrology* No. 48, UNESCO, Paris, 2001.

Barrett, Curtis, "Development of Global Integrated Water Management Systems", *Journal of Management in Engineering*, American Society of Civil Engineers, Vol. 15,No. 4, July/August, 1999

Enarson, Elaine,"Gender and Natural Disasters", *Working Paper I*, International Labour Organization (ILO)/Recovery and Reconstruction Department, Geneva, September, 2000.

_____________ , "Gender Equality, Environmental Management, and Natural Disaster Mitigation", *Report from the On-Line Conference conducted by the United Nations*, Department for Economic and Social Affairs, Division for the Advancement of Women, New York, November, 2001.

_____________ and Morrow, Betty Hearn (eds.), "The Gendered Terrain of Disaster: Through Women's Eyes", Greenwood Publishing, 1998.

Freeman, Paul K.; Martin, Leslie A.; Mechler, Reinhard and Warner, Koko (eds.), "Catastrophes and Development. Integrating Natural Catastrophes into Development Planning", *Disaster Risk Management Working Paper Series* No. 4, World Bank, Washington, D.C., April, 2002.

Gilbert, Roy and Kreimer, Alcira, "Learning from World Bank's Experience of Natural Disasters Related Assistance", *World Bank Paper Series No. 2*, Disaster Management Facility, World Bank, Washington, D.C., 1999.

Parker, D.J. (ed.), "Floods Vol. I & Vol. II ", *Routledge Hazards and Disasters Series*, Routledge, London, 2000.

United Nations Economic and Social Commission for Asia and the Pacific, "Regional Cooperation in the Twenty-First Century on Flood Control and Management in Asia and the Pacific", United Nations: New York, 1999.

United Nations Department of Economic and Social Affairs, "Natural Disasters and Sustainable Development: Understanding the links between Development, Environment and Natural Disasters", *Background,* 2002.

World Bank, "Managing Economic Crisis and Natural Disasters", in *World Development Report 2000/2001,* (Chap. 9), World Bank, Washington, D.C., 2001.

World Meteorological Organization (WMO), "Comprehensive Risk Assessment for Natural Hazards", WMO/TD No. 955, Geneva, 1999.

Index

Notes

Notes

Notes

Notes